Random measures

RANDOM MEASURES

Olav Kallenberg
Department of Mathematics
University of Göteborg

1976

AKADEMIE-VERLAG · BERLIN

and

ACADEMIC PRESS · LONDON — NEW YORK — SAN FRANCISCO

Erschienen im Akademie-Verlag, 108 Berlin, Leipziger Straße 3—4

Lizenznummer: 202 · 100/556/76
Bestellnummer: 762 423 2 (6414) · LSV 1075
ACADEMIC PRESS INC. (LONDON) LTD.
24/28 Oval Road
London NW 1
United States Edition published by
ACADEMIC PRESS INC.
111 Fifth Avenue
New York, New York 10 003
Library of Congress Catalog Card Number: 76-20 435
ISBN: 012-394 950-5
Printed in GDR

Contents

To my mother

Introduction and preface

Like most branches of probability, the theory of random measures has its origin in applications. Thus PALM [65] made in 1943 an extensive study of point processes on the line as models for telephone traffic, and many basic ideas of point process theory are due to him. PALM's ideas were developed further and made rigorous by HINČIN [24] in his classical work on queuing theory.

While point processes on the line could easily be described and analysed in terms of the corresponding step processes, a similar approach would be inconvenient in higher dimensions, and so the step into the multidimensional case, taken by DOOB [14], p. 405, in connection with infinite particle systems (see also [12]), was an important one.

Not until around 1960, point processes on the line attracted interest for their own sake, important progress then being made by RÉNYI [70, 71], RYLL-NARDZEWSKI [72], SLIVNYAK [74], NAWROTZKI [62], GRIGELIONIS [19], KERSTAN, MATTHES [41, 42, 43, 54] and others. However, the modern era began with the work of PROHOROV [69], JIŘINA [29], MECKE [56], WALDENFELS [76] and HARRIS [22, 23], who cleared the path for an extension of the whole theory to the general setting of random measures on abstract spaces, this turning out to be possible essentially without any additional effort. The abstract point of view introduced by these authors has proved extremely fruitful even for the study of point processes on the line, partly because it helps to concentrate on the topological structure of the space (which has proved important) and to dispense with its order structure (which has usually turned out to be irrelevant). As a result, the bulk of knowledge is now increasing rapidly, due to the efforts of innumerable scolars in DDR and other countries. An extensive account of the state of point process theory around 1972 is given by KERSTAN, MATTHES and MECKE [44].

The present aim is to give a systematic account of some basic ideas of general random measure theory. As indicated above, the latter appears to be the natural scope of a theory, at least from a mathematical point of view. Indeed, most results of point process theory carry over to this more general context, while a further extension to general random processes is usually impossible. Furthermore, general random measures are needed for a description of many point process results, (see e.g. Theorem 8.4 below). Finally, random measure theory appears to be the natural tool in many models involving non-negative random quantities, and here point processes are appropriate only in the particular case when these quantities are integer valued, (cf. [33] or Theorem 9.4 below).

Most material is developed from my own work in [31, 32, 36, 37, 38], and the approach is essentially the same. However, the theory is extended and improved throughout, and a large number of new results are included. Moreover, complete proofs are now given even in cases where only scetchy arguments were available before. To facilitate the access of the reader, I have further included, in an appendix, some basic facts from topology, measure theory and probability, thus making the exposition reasonably self-contained.

In order not to drown the reader into details, only the basic facts are included in the main text, while variants and extensions are given as exercises. It should be emphasised, however, that the latter constitute an equally important part of the work, containing in particular many new results. Naturally, I have had no intention to attain completeness. For example, the important notion of stationarity is not even mentioned (except in exercises), mainly because I have nothing essential to add to the exhaustive treatment in [44], Chapter 3.

The volume is a revised and completed version of a series of lectures which I gave at the Statistics Department, Chapel Hill, in the spring semester of 1974, and during this period, my work was supported in part by the Office of Naval Research under Contract N00014-67-A-0321-002 with the University of North Carolina. Several persons have contributed to the final result by criticizing early versions of the manuscript, and I wish to thank them all. In particular, I would like to mention professor M. R. LEADBETTER for some helpful comments and for his constant encouragement. I am further most grateful to professor K. MATTHES in Berlin for his stimulating interest and valuable advice.

The final version of the manuscript was prepared in Göteborg in January 1975.

OLAV KALLENBERG

and note that $\mathscr{D}$ is closed under bounded monotone limits. Furthermore, $\mathscr{D}$ contains the ring $\mathscr{C}$ of all finite unions of $\mathscr{I}$-sets, since every such union may be taken to be disjoint. Hence we may conclude from A 2.2 that $\mathscr{D} \supset \hat{\sigma}(\mathscr{C}) = \hat{\sigma}(\mathscr{I}) = \mathscr{B}$, which yields the desired measurability of $\mu \to \mu B$, $B \in \mathscr{B}$.

We next consider the σ-algebra $\mathscr{M}''$ generated by all mappings $\mu \to \mu f$, $f \in \mathscr{F}_c$, and note that $\mathscr{M}'' \subset \mathscr{M}$ by Lemma 1.3. To prove the converse relation, let $C \subset \mathfrak{S}$ be compact, and choose a sequence $f_1, f_2, \ldots \in \mathscr{F}_c$ satisfying $f_n \downarrow 1_C$, (cf. A6.1). Then $\mu f_n \downarrow \mu C$, $\mu \in \mathfrak{M}$, by dominated convergence, which proves the $\mathscr{M}''$-measurability of $\mu \to \mu C$. We may now complete the proof as in case of Lemma 1.2, defining for fixed compact C

$$\mathscr{D} = \{B \in \mathscr{S} : \mu \to \mu(B \cap C) \text{ is } \mathscr{M}''\text{-measurable}\} . \qquad \square$$

Lemma 1.5. $\mathscr{N} \subset \mathscr{M}$.

Proof. By the definitions of $\mathscr{M}$ and $\mathscr{N}$ we have $\mathscr{N} \subset \mathscr{M} \cap \mathfrak{N} = \{M \cap \mathfrak{N} : M \in \mathscr{M}\}$, so it is enough to prove that $\mathfrak{N} \in \mathscr{M}$. Let $\mathscr{U} \subset \mathscr{B}$ be an arbitrary countable DC-ring, and define

$$M = \{\mu \in \mathfrak{M} : \mu U \in Z_+,\ U \in \mathscr{U}\} .$$

We intend to show that $\mathfrak{N} = M$; since clearly $M \in \mathscr{M}$, this will complete the proof. Now $\mathfrak{N} \subset M$ holds trivially, so it remains to prove that $M \subset \mathfrak{N}$. For this purpose, define for fixed $\mu \in \mathfrak{M}$

$$\mathscr{D} = \{B \in \mathscr{B} : \mu B \in Z_+\} .$$

Since $\mathscr{D}$ is closed under bounded monotone limits and contains the ring $\mathscr{C}$ generated by $\mathscr{I}$, it is seen from A 2.2 and Lemma 1.2 that $\mathscr{D} \supset \hat{\sigma}(\mathscr{C}) = \hat{\sigma}(\mathscr{I}) = \mathscr{B}$. Thus $\mu \in \mathfrak{N}$, so we have indeed $M \subset \mathfrak{N}$. $\square$

Random measures and point processes

By a *random measure* or a *point process* on $\mathfrak{S}$ we mean any measurable mapping of some fixed probability space $(\Omega, \mathscr{A}, P)$ into $(\mathfrak{M}, \mathscr{M})$ or $(\mathfrak{N}, \mathscr{N})$ respectively. By Lemma 1.4, a point process may alternatively be considered as an $\mathfrak{N}$-valued random measure, and conversely any a.s. $\mathfrak{N}$-valued random measure coincides a.s. with a point process. Thus we shall make no difference in the sequel between point processes and a.s. $\mathfrak{N}$-valued random measures. Similarly, we shall allow a random measure to take values outside $\mathfrak{M}$ on a set $A \in \mathscr{A}$ with $PA = 0$.

Lemma 1.6. *The class of random measures (or point processes) on $\mathfrak{S}$ is closed under addition and under multiplication by R_+-valued (or Z_+-valued respectively) random variables. Furthermore, a series $\sum_j \xi_j$ of random measures (or point processes) is itself a random measure (or point process) iff $\sum_j \xi_j B < \infty$ a.s. for all $B \in \mathscr{B}$.*

Proof. The first assertion follows immediately from the definition of $\mathscr{M}(\mathscr{N})$ and the fact that the class of random variables is closed under addition and multiplication. As for the second assertion, it is seen by monotone convergence that $\sum_j \xi_j$ is σ-additive on $\mathscr{B}$, and hence measure valued. Moreover, the necessity of our condition follows from the fact that $\xi B < \infty$ a.s. for any random measure ξ and any $B \in \mathscr{B}$. Suppose conversely that $\sum_j \xi_j B < \infty$ a.s., $B \in \mathscr{B}$. Considering this inequality for all sets B belonging to some countable covering class, it is seen that the exceptional P-null set may be taken to be independent of B. Thus $\sum_j \xi_j \in \mathfrak{M}\,(\mathfrak{N})$ a.s. Finally, the measurability of $\sum_j \xi_j$ follows from the fact that $\sum_j \xi_j B$ is a random variable for every $B \in \mathscr{B}$. □

The *distribution* of a random measure or point process ξ is by definition the probability measure $P\xi^{-1}$ on $(\mathfrak{M}, \mathscr{M})$ or $(\mathfrak{N}, \mathscr{N})$ given by

$$(P\xi^{-1})\,M = P(\xi^{-1}M) = P\{\xi \in M\}\,, \qquad M \in \mathscr{M} \text{ or } \mathscr{N}\,.$$

We further define the *intensity* $E\xi$ of ξ as the set function

$$(E\xi)B = E(\xi B)\,, \qquad B \in \mathscr{B}\,,$$

where E denotes the expectation or integral w.r.t. P. Note that $E\xi$ is always a measure, though it need not belong to $\mathfrak{M}$. We finally define the L-*transform* (L for LAPLACE) L_ξ of ξ by

$$L_\xi(f) = Ee^{-\xi f}\,, \qquad f \in \mathscr{F}\,.$$

Note in particular that, for any $k \in N$ and $B_1, \ldots, B_k \in \mathscr{B}$, the function

$$L_\xi\Big(\sum_j t_j 1_{B_j}\Big) = E\exp\Big(-\sum_j t_j \xi B_j\Big) = L_{\xi B_1, \ldots, \xi B_k}(t_1, \ldots, t_k)\,, \qquad t_1, \ldots, t_k \in R_+\,,$$

is the elementary L-transform of the random vector $(\xi B_1, \ldots, \xi B_k)$.

Many important random measure distributions are defined most easily by means of mixing. In the present context, the general measurability requirement A 3.2 can be relaxed as follows.

Lemma 1.7. *Let $(\mathfrak{T}, \mathscr{T}, Q)$ be a probability space and let ξ_ϑ, $\vartheta \in \mathfrak{T}$, be a family of random measures on $\mathfrak{S}$. Then the mixture of $P\xi_\vartheta^{-1}$ w.r.t. Q exists iff $L_{\xi_\vartheta}(f)$ $\mathscr{T}$-measurable in ϑ for every $f \in \mathscr{F}_c$.*

Proof. By A 3.2, the mixture exists iff $P\{\xi_\vartheta \in M\}$ is $\mathscr{T}$-measurable for every $M \in \mathscr{M}$, so let us first assume that this condition is fulfilled. Then $P\{\xi_\vartheta f \leqq x\}$ is $\mathscr{T}$-measurable for every $f \in \mathscr{F}_c$ and $x \in R_+$ according to Lemma 1.3, and it follows by the definition of the integral that $L_{\xi_\vartheta}(f) = Ee^{-\xi_\vartheta f}$ is measurable for every $f \in \mathscr{F}_c$.

Conversely, suppose that $L_{\xi_\vartheta}(f)$ is $\mathscr{T}$-measurable for every $f \in \mathscr{F}_c$. Then so is

$$L_{\xi_\vartheta f_1, \ldots, \xi_\vartheta f_k}(t_1, \ldots, t_k) = L_{\xi_\vartheta}\Big(\sum_j t_j f_j\Big)$$

for every $k \in N$, $f_1, \ldots, f_k \in \mathscr{F}_c$ and $t_1, \ldots, t_k \in R_+$, and scrutinizing the proof of the uniqueness theorem A5.1 for multidimensional L-transforms, it is seen that $P\{\xi_\vartheta f_1 \leqq x_1, \ldots, \xi_\vartheta f_k \leqq x_k\}$ is $\mathscr{T}$-measurable for any $k \in N$, $f_1, \ldots, f_k \in \mathscr{F}_c$ and $x_1, \ldots, x_k \in R_+$. We now introduce the class $\mathscr{D}$ of all sets $M \in \mathscr{M}$ such that $P\{\xi_\vartheta \in M\}$ is $\mathscr{T}$-measurable, and note that $\mathscr{D}$ contains $\mathscr{M}$ and is closed under proper differences and monotone limits. Furthermore, it was seen above that $\mathscr{D}$ contains the class $\mathscr{C}$ of all sets of the form

$$\{\mu \in \mathscr{M}: \mu f_1 \leqq x_1, \ldots, \mu f_k \leqq x_k\}\,, \quad k \in N\,, \quad f_1, \ldots, f_k \in \mathscr{F}_c\,, \quad x_1, \ldots, x_k \in R_+\,,$$

and the latter class being closed under finite intersections, it follows by A2.1 that $\mathscr{D} \supset \sigma(\mathscr{C})$. Since $\sigma(\mathscr{C}) = \mathscr{M}$ by Lemma 1.4, this means that $P\{\xi_\vartheta \in M\}$ is indeed $\mathscr{T}$-measurable for every $M \in \mathscr{M}$. □

Examples

For any fixed $s \in \mathfrak{S}$, we define the DIRAC *measure* $\delta_s \in \mathfrak{N}$ by $\delta_s B = 1_B(s)$, $B \in \mathscr{B}$. The mapping $s \to \delta_s$ is clearly measurable $(\mathfrak{S}, \mathscr{S}) \to (\mathfrak{N}, \mathscr{N})$, and in particular δ_τ is a point process on $\mathfrak{S}$ for any random element τ in $(\mathfrak{S}, \mathscr{S})$. Writing $\omega = P\tau^{-1}$, it is seen that δ_τ has intensity $E\delta_\tau = P\tau^{-1} = \omega$ and L-transform

$$Ee^{-\delta_\tau f} = Ee^{-f(\tau)} = \omega e^{-f}, \qquad f \in \mathscr{F}. \tag{1.1}$$

Next suppose that $n \in Z_+$ and let $\tau_1, \ldots, \tau_n$ be independent random elements in $\mathfrak{S}$ with common distribution ω. We shall say that a point process ξ on $\mathfrak{S}$ is a *sample process* with intensity $n\omega$, if ξ has the same distribution as $\delta_{\tau_1} + \cdots + \delta_{\tau_n}$ By (1.1) and the assumed independence, ξ has then the L-transform

$$E \exp\left(-\sum_{j=1}^{n} \delta_{\tau_j} f\right) = \prod_{j=1}^{n} E \exp(-\delta_{\tau_j} f) = (\omega e^{-f})^n, \qquad f \in \mathscr{F}. \tag{1.2}$$

By Lemma 1.7, we may consider $n = \nu$ as a Z_+-valued random variable and mix w.r.t. its distribution to obtain a *mixed sample process* with intensity $(E\nu)\omega$ and L-transform

$$Ee^{-\xi f} = E(\omega e^{-f})^\nu = \psi(\omega e^{-f}), \qquad f \in \mathscr{F}. \tag{1.3}$$

Here ψ denotes the (probability) *generating function* of ν, i.e.

$$\psi(s) = Es^\nu, \qquad s \in [0, 1].$$

In the particular case when ν is POISSONian with mean $a \geqq 0$, ψ is given by

$$\psi(s) = e^{-a} \sum_{n=0}^{\infty} \frac{a^n s^n}{n!} = e^{-a(1-s)}, \qquad s \in [0, 1],$$

and (1.3) becomes, with $\lambda = a\omega$,

$$Ee^{-\xi f} = e^{-a(1-\omega e^{-f})} = e^{-\lambda(1-e^{-f})}, \qquad f \in \mathscr{F}, \tag{1.4}$$

where we have used the fact that $\omega\mathfrak{S} = 1$. A point process with this distribution is called a POISSON *process* with intensity λ, In this case λ is bounded, but we may also construct POISSON processes with unbounded intensity $\lambda \in \mathfrak{M}$. For this purpose, let $S_1, S_2, \ldots \in \mathscr{B}$ be any disjoint partition of $\mathfrak{S}$ into bounded sets, (cf. A6.1 for existence). Since the corresponding restrictions $S_j\lambda$, $j \in N$, of λ are bounded, there exist some independent POISSON processes $\xi_1, \xi_2, \ldots$ on $\mathfrak{S}$ with these measures as intensities. Moreover,

$$\sum_j E\xi_j B = \sum_j (S_j\lambda) B = \lambda B < \infty, \qquad B \in B,$$

so the series $\sum \xi_j$ converges by Lemma 1.6 to some point process ξ on $\mathfrak{S}$. Applying (1.4) to each ξ_j and using the assumed independence, we obtain for any $f \in \mathscr{F}$

$$Ee^{-\xi f} = \prod_j Ee^{-\xi_j f} = \prod_j \exp\{-(S_j\lambda)(1 - e^{-f})\}$$
$$= \exp\left\{-\sum_j (S_j\lambda)(1 - e^{-f})\right\} = e^{-\lambda(1-e^{-f})},$$

so (1.4) remains true. (The formal calculations here and in similar places are justified by the fact that all quantities involved are non-negative.) Any point process ξ with L-transform $e^{-\lambda(1-e^{-f})}$ will henceforth be called a POISSON process with intensity λ. As will be seen from Theorem 3.1, its distribution $P\xi^{-1}$ is uniquely determined by λ.

If ξ is a POISSON process on $\mathfrak{S}$ with intensity λ and if $B_1, \ldots, B_k \in \mathscr{B}$ are disjoint, we get for any $t_1, \ldots, t_k \in R_+$

$$L_{\xi B_1,\ldots,\xi B_k}(t_1, \ldots, t_k) = L_\xi\Big(\sum_j t_j 1_{B_j}\Big) = \exp\Big\{-\lambda\Big[1 - \exp\Big(-\sum_j t_j 1_{B_j}\Big)\Big]\Big\}$$
$$= \exp\Big\{-\lambda \sum_j 1_{B_j}(1 - e^{-t_j})\Big\} = \prod_j \exp\{-\lambda 1_{B_j}(1 - e^{-t_j})\}$$
$$= \prod_j L_\xi(t_j 1_{B_j}) = \prod_f L_{\xi B_j}(t_j)\ .$$

By A5.1, this shows that ξ has *independent increments*, in the sense that $\xi B_1, \ldots, \xi B_k$ are independent for any $k \in N$ and disjoint $B_1, \ldots, B_k \in \mathscr{B}$. This fact was actually the basis for the above construction of POISSON processes with unbounded intensity.

Let us next consider a POISSON process ξ_α with intensity $\alpha\lambda$, where $\alpha \in R_+$ while $\lambda \in \mathfrak{M}$. By (1.4) and Lemma 1.7, we may consider α as an R_+-valued random variable and mix w.r.t. its distribution, thus obtaining a *mixed* POISSON *process*, possessing the L-transform

$$Ee^{-\alpha\lambda(1-e^{-f})} = L_\alpha\big(\lambda(1 - e^{-f})\big)\ , \qquad f \in \mathscr{F}\ . \tag{1.5}$$

More generally, it is seen from (1.4) and Lemmas 1.3 and 1.7 that the intensity $\lambda = \eta$ of a POISSON process may be considered as a random measure on $\mathfrak{S}$. In this way we obtain by mixing a COX *process* ξ *directed by* η, possessing the L-transform

$$L_\xi(f) = Ee^{-\eta(1-e^{-f})} = L_\eta(1 - e^{-f})\ , \qquad f \in \mathscr{F}\ . \tag{1.6}$$

Table 1. Some basic point processes and random measures.

process ξ	based on	$E\xi$	$P\{\xi B = 0\}$	$L_\xi(f)$
sample	ω, n	$n\omega/\omega\mathfrak{S}$	$(\omega B^c/\omega\mathfrak{S})^n$	$(\omega e^{-f}/\omega\mathfrak{S})^n$
mixed sample	$\omega, \nu\ (\psi)$	$(E\nu)\ \omega/\omega\mathfrak{S}$	$\psi(\omega B^c/\omega\mathfrak{S})$	$\psi(\omega e^{-f}/\omega\mathfrak{S})$
POISSON	λ	λ	$e^{-\lambda B}$	$e^{-\lambda(1-e^{-f})}$
mixed POISSON	λ, α	$(E\alpha)\ \lambda$	$L_\alpha(\lambda B)$	$L_\alpha(\lambda(1 - e^{-f}))$
COX	η	$E\eta$	$Ee^{-\eta B}$	$L_\eta(1 - e^{-f})$
compound	η, β	$(E\beta)\ E\eta$	$E(P\{\beta = 0\})^{\eta B}$	$L_\eta(-\log L_\beta \circ f)$
thinning	η, p	$pE\eta$	$E(1 - p)^{\eta B}$	$L_\eta(-\log[1 - p(1-e^{-f})])$

We finally introduce compound point processes as follows. As will be seen in Lemma 2.1 below, there exists for every fixed $\mu \in \mathfrak{N}$ some finite or infinite sequence $t_1, t_2, \ldots \in \mathfrak{S}$ such that $\mu = \sum \delta_{t_j}$. Assuming $\beta, \beta_1, \beta_2, \ldots$ to be independent and identically distributed R_+-valued random variables, it follows by Lemma 1.6 that $\xi = \sum \beta_j \delta_{t_j}$ is a random measure on $\mathfrak{S}$, and by the assumed independence, its L-transform becomes for $f \in \mathscr{F}$ (writing $\circ$ for composition)

$$E\exp\Big(-\sum_j \beta_j f(t_j)\Big) = \prod_j E\exp\big(-\beta f(t_j)\big) = \prod_j L_\beta \circ f(t_j)$$
$$= \exp\sum_j \log L_\beta \circ f(t_j) = \exp(\mu \log L_\beta \circ f)\ .$$

According to Lemmas 1.3 and 1.7, we may mix here w.r.t. $\mu = \eta$ when regarded as a point process on $\mathfrak{S}$, and in this way we obtain a β-*compound of* η, possessing

Proof. Put for brevity $\xi_a^* - \xi_b^* = \xi_{[a,b)}^*$. Taking differences in Lemma 2.2, it is seen that

$$\lim_{n\to\infty} \sum_j 1_{[a,b)}(\xi B_{nj}) = \xi_{[a,b)}^* B , \tag{2.9}$$

and hence by FATOU's lemma

$$E\xi_{[a,b)}^* B \leqq \liminf_{n\to\infty} E \sum_j 1_{[a,b)}(\xi B_{nj}) = \liminf_{n\to\infty} \sum_j P\{a \leqq \xi B_{nj} < b\} ,$$

which proves (2.8) in the case when $E\xi_{[a,b)}^* B = \infty$. Thus it remains to consider the case when $E\xi_a' B$ and $E\xi_{[a,b)}^* B$ are both finite. But then (2.8) follows from (2.9) by dominated convergence, since

$$\{a \leqq \xi B_{nj} < b\} \subset \{\xi_{[a,b)}^* B_{nj} \geqq 1\} \cup \{\xi_a' B_{nj} \geqq a\} ,$$

and therefore

$$\begin{aligned}\sum_j 1_{[a,b)}(\xi B_{nj}) &\leqq \sum_j 1_{[1,\infty)}(\xi_{[a,b)}^* B_{nj}) \vee 1_{[a,\infty)}(\xi_a' B_{nj}) \\ &\leqq \sum_j 1_{[1,\infty)}(\xi_{[a,b)}^* B_{nj}) + \sum_j 1_{[a,\infty)}(\xi_a' B_{nj}) \\ &\leqq \sum_j \xi_{[a,b)}^* B_{nj} + \sum_j a^{-1} \xi_a' B_{nj} = \xi_{[a,b)}^* B + a^{-1} \xi_a' B . \qquad \square\end{aligned}$$

Given any covering class $\mathcal{J} \subset \mathcal{B}$ and any $a > 0$, we shall say that ξ is *a-regular* w.r.t. $\mathcal{J}$, if for every fixed $I \in \mathcal{J}$ there exists some array $\{I_{nj}\} \subset \mathcal{J}$ of finite covers of I (one for each $n \in N$) such that

$$\lim_{n\to\infty} \sum_j P\{\xi I_{nj} \geqq a\} = 0 . \tag{2.10}$$

Theorem 2.5. *Let ξ be a random measure on $\mathfrak{S}$, let $\mathcal{J} \subset \mathcal{B}$ be a covering class and let $a > 0$. Then $\xi_a^* = 0$ a.s. provided ξ is a-regular w.r.t. $\mathcal{J}$. The converse is also true if $E\xi \in \mathfrak{M}$ and $\mathcal{J}$ is a DC-semiring.*

Proof. Let $I \in \mathcal{J}$ be fixed and let $\{I_{nj}\} \subset \mathcal{J}$ be an array of finite covers of I satisfying (2.10). Then

$$P\{\xi_a^* I > 0\} \leqq P \bigcup_j \{\xi I_{nj} \geqq a\} \leqq \sum_j P\{\xi I_{nj} \geqq a\} \to 0 ,$$

which yields $\xi_a^* I = 0$ a.s. Thus the first assertion follows from the fact that $\mathcal{J}$ is covering. Conversely, suppose that $\xi_a^* = 0$ a.s. and $E\xi \in \mathfrak{M}$, and let $\mathcal{J}$ be a DC-semiring. Choose for any fixed $I \in \mathcal{J}$ some null-array $\{I_{nj}\} \subset \mathcal{J}$ of partitions of I, and conclude from Theorem 2.4 that

$$\lim_{n\to\infty} \sum_j P\{\xi I_{nj} \geqq a\} = E\xi_n^* I = 0 .$$

Since I was arbitrary, this proves the asserted regularity of ξ. $\square$

It is sometimes useful to replace the global regularity condition (2.10) by a local one. For a first step, note that if $\mathcal{J}$ is a DC-semiring and if $\lambda \in \mathfrak{M}$, then (2.10) holds if $P\{\xi I \geqq a\} = o(\lambda I)$ as $|I| \to 0$, uniformly for all $I \in \mathcal{J}$ contained in an arbitrary compact set, or equivalently, if

$$\lim_{\varepsilon\to 0} \sup \left\{\frac{1}{\lambda I} P\{\xi I \geqq a\} : I \in \mathcal{J} \cap I_0, |I| < \varepsilon\right\}, \qquad I_0 \in \mathcal{J} , \tag{2.11}$$

so in this case $\xi_a^* = 0$ a.s. (Here and below, 0/0 is to be interpreted as 0.) We shall show that the same conclusion may be drawn even without the uniformity requirement. Let us say that a sequence of partitions is *nested* if it proceeds by successive refinements.

Theorem 2.6. *Let ξ be a random measure on $\mathfrak{S}$ and let $a > 0$. Then $\xi_a^* = 0$ a.s., provided there exist some DC-semiring $\mathcal{J} \subset \mathcal{B}$ and some $\lambda \in \mathfrak{M}$ satisfying*

$$\limsup_{\varepsilon \to 0} \left\{ \frac{1}{\lambda I} P\{\xi I \geqq a\} : I \in \mathcal{J},\, |I| < \varepsilon,\, s \in I^- \right\} = 0\,, \qquad s \in \mathfrak{S}\,. \tag{2.12}$$

Proof. Let $\mathcal{J}$ and λ be such as stated, and suppose that $P\{\xi_a^* \neq 0\} > 0$. Since $\mathcal{J}$ is a covering class, we may then choose some set $I \in \mathcal{J}$ such that $P\{\xi_a^* I > 0\} > 0$. Letting $\{I_{nj}\} \subset \mathcal{J}$ be a null-array of nested partitions of I, we obtain

$$P\{\xi_a^* I > 0\} = P \bigcup_j \{\xi_a^* I_{1j} > 0\} \leqq \sum_j P\{\xi_a^* I_{1j} > 0\}\,, \tag{2.13}$$

and we shall prove that this implies

$$\max_j \frac{1}{\lambda I_{1j}} P\{\xi_a^* I_{1j} > 0\} \geqq \frac{1}{\lambda I} P\{\xi_a^* I > 0\}\,. \tag{2.14}$$

In fact, (2.14) follows trivially from (2.13) if $\lambda I = 0$, while if $\lambda I > 0$, insertion of the converse of (2.14) into (2.13) would yield the contradiction

$$P\{\xi_a^* I > 0\} < \frac{1}{\lambda I} P\{\xi_a^* I > 0\} \sum_j \lambda I_{1j} = P\{\xi_a^* I > 0\}\,.$$

From (2.14) it is seen that there exists some $I_1 \in \{I_{11}, I_{12}, \ldots\}$ satisfying

$$\frac{1}{\lambda I_1} P\{\xi_a^* I_1 > 0\} \geqq \frac{1}{\lambda I} P\{\xi_a^* I > 0\}\,,$$

and proceeding inductively, we may construct a non-increasing sequence $I_1, I_2, \ldots \in \mathcal{J}$ with $|I_n| \to 0$, such that

$$\frac{1}{\lambda I_n} P\{\xi I_n \geqq a\} \geqq \frac{1}{\lambda I_n} P\{\xi_a^* I_n > 0\} \geqq \frac{1}{\lambda I} P\{\xi_a^* I > 0\} > 0\,, \quad n \in N\,. \tag{2.15}$$

Since $\{I_n^-\}$ has a non-empty intersection $\{s\}$, (2.15) shows that (2.12) is violated at s. □

Criteria for a.s. diffuseness of random measures are obtained from Theorems 2.5 and 2.6 by letting a be arbitrary > 0. Thus ξ is a.s. diffuse if ξ is a-regular w.r.t. some covering class $\mathcal{J} \subset \mathcal{B}$ for every $a > 0$. In this case we shall say that ξ is *regular* w.r.t. $\mathcal{J}$. Note also that simplicity criteria for point processes may be obtained by taking $a = 2$ in Theorems 2.5 and 2.6. Finally observe that, for point processes, (2.8) is generally true with $a = 1$, since in this case $\xi_1' = 0$ a.s.

We conclude this section with a different kind of simplicity criterion which will play an important role in the sequel.

Lemma 2.7. *Let ξ be a point process on $\mathfrak{S}$ and let $\mathcal{J} \subset \mathcal{B}$ be a DC-semiring. Then ξ is a.s. simple iff*

$$P\{\xi I > 1\} \leqq P\{\xi^* I > 1\}\,, \qquad I \in \mathcal{J}\,. \tag{2.16}$$

Proof. Suppose that (2.16) holds, and conclude that

$$0 \leqq P\{\xi I > \xi^* I = 1\} = P\{\xi I > 1\} - P\{\xi^* I > 1\} \leqq 0\,, \qquad I \in \mathcal{J}\,. \tag{2.17}$$

Since $\mathscr{C}$ is closed under finite intersections, we may conclude from A 2.1 that $\mathscr{D} \supset \sigma(\mathscr{C})$. But $\sigma(\mathscr{C}) = \mathscr{M}$ by Lemma 1.4, and so $P\{\xi \in M\} = P\{\eta \in M\}$ for all $M \in \mathscr{M}$, i.e. $\xi \stackrel{d}{=} \eta$.

Let us next assume that (ii)′ holds. Then

$$L_{\xi f_1, \ldots, \xi f_k}(t_1, \ldots, t_k) = L_\xi\Big(\sum_j t_j f_j\Big) = L_\eta\Big(\sum_j t_j f_j\Big) = L_{\eta f_1, \ldots, \eta f_k}(t_1, \ldots, t_k)$$

for every $k \in N$, $f_1, \ldots, f_k \in \mathscr{F}_c$ and $t_1, \ldots, t_k \in R_+$, so it follows by A5.1 that

$$(\xi f_1, \ldots, \xi f_k) \stackrel{d}{=} (\eta f_1, \ldots, \eta f_k), \qquad k \in N, \qquad f_1, \ldots, f_k \in \mathscr{F}_c.$$

We may now proceed as in the first part of the proof to conclude that (i) holds. Since (ii) trivially implies (ii)′, this completes the proof. □

Corollary 3.2. *Let ξ be a Cox process directed by some random measure η on $\mathfrak{S}$. Then the distributions of ξ and η determine each other uniquely. This is also true when ξ is a β-compound of some point process η, provided $P\beta^{-1}$ is known and such that $\beta \stackrel{d}{\neq} 0$.*

Proof. If ξ is a Cox process directed by η, we have by Table 1

$$L_\xi(f) = L_\eta(1 - e^{-f}), \qquad f \in \mathscr{F}. \tag{3.1}$$

Writing $1 - e^{-f} = tg$, we may solve for f provided $0 \leqq tg < 1$, thus obtaining $f = -\log(1 - tg)$. Hence by (3.1)

$$L_{\eta g}(t) = L_\eta(tg) = L_\xi\big(-\log(1 - tg)\big), \qquad 0 \leqq t < ||g||^{-1}, \qquad g \in \mathscr{F}_c,$$

where $||g|| = \sup_t g(t)$, and so it is seen from A 5.1 that $P\xi^{-1}$ determines $P(\eta g)^{-1}$ for all $g \in \mathscr{F}_c$. But by Theorem 3.1, the latter distributions determine $P\eta^{-1}$.

In case of β-compounds, we get in place of (3.1)

$$L_\xi(f) = L_\eta(-\log L_\beta \circ f), \qquad f \in \mathscr{F}, \tag{3.2}$$

where the ring stands for composition of functions. Putting $-\log L_\beta \circ f = tg$ and noting that L_β has a unique inverse L_β^{-1} on the interval $(P\{\beta = 0\}, 1]$, it is seen that we may solve for f provided $0 < tg < -\log P\{\beta = 0\}$, thus obtaining $f = L_\beta^{-1} \circ e^{-tg}$. Hence by (3.2)

$$L_{\eta g}(t) = L_\xi(L_\beta^{-1} \circ e^{-tg}), \qquad 0 \leqq t < -||g||^{-1} \log P\{\beta = 0\}, \qquad g \in \mathscr{F}_c,$$

and since $P\{\beta = 0\} < 1$, the proof may be completed as before. □

Simplicity and diffuseness

Theorem 3.1 may be partially strengthened as follows.

Theorem 3.3. *Let ξ and η be point processes on $\mathfrak{S}$, and suppose that ξ is a.s. simple. Further suppose that $\mathscr{U} \subset \mathscr{B}$ is a DC-ring while $\mathscr{I} \subset \mathscr{B}$ is a DC-semiring. Then $\xi \stackrel{d}{=} \eta^*$ iff*

$$P\{\xi U = 0\} = P\{\eta U = 0\}, \qquad U \in \mathscr{U}. \tag{3.3}$$

Furthermore, $\xi \stackrel{d}{=} \eta$ iff (3.3) holds and in addition

$$P\{\xi I > 1\} \geqq P\{\eta I > 1\}, \qquad I \in \mathscr{I}. \tag{3.4}$$

Proof. The necessity assertions are obvious. Suppose conversely that (3.3) holds, and define

$$\mathcal{D} = \{M \in \mathcal{N}: P\{\xi \in M\} = P\{\eta \in M\}\} .$$

Then $\mathcal{D}$ contains $\mathfrak{N}$ and is closed under proper differences and monotone limits. Furthermore, it follows from (3.3) that $\mathcal{D}$ contains the class

$$\mathcal{C} = \{\{\mu \in \mathfrak{N}: \mu U = 0\} , \qquad U \in \mathcal{U}\} .$$

Now $\mathcal{C}$ is closed under finite intersections, since $\mathcal{U}$ is closed under finite unions and moreover

$$\{\mu U_1 = 0\} \cap \{\mu U_2 = 0\} = \{\mu(U_1 \cup U_2) = 0\} , \qquad U_1, U_2 \in \mathcal{U} ,$$

so it follows by A 2.1 that $\mathcal{D} \supset \sigma(\mathcal{C})$. Using Lemma 2.2, it is further seen that the mapping $\mu \to \mu^* I$ is $\sigma(\mathcal{C})$-measurable for every $I \in \mathcal{J}$, and by Lemmas 1.3 and 1.4 this proves that the mapping φ: $\mu \to \mu^*$ is measurable $\sigma(\mathcal{C}) \to \mathcal{N}$. Using the fact that $\mathcal{D} \supset \sigma(\mathcal{C})$, we thus obtain

$$P\{\xi^* \in M\} = P\{\xi \in \varphi^{-1}M\} = P\{\eta \in \varphi^{-1}M\} = P\{\eta^* \in M\} , \quad M \in \mathcal{N},$$

which proves that $\xi^* \stackrel{d}{=} \eta^*$, and hence also that $\xi \stackrel{d}{=} \eta^*$.

If (3.4) holds in addition, then

$$P\{\eta^* I > 1\} = P\{\xi I > 1\} \geqq P\{\eta I > 1\} , \qquad I \in \mathcal{J} ,$$

and it follows by Lemma 2.7 that η is a.s. simple. Hence $\xi \stackrel{d}{=} \eta$ in this case. □

Theorem 3.4. *Let ξ and η be point processes (or random measures) on $\mathfrak{S}$, and suppose that ξ is a.s. simple (or diffuse respectively). Further suppose that $\mathcal{U} \subset \mathcal{B}$ is a DC-ring while $\mathcal{C} \subset \mathcal{B}$ is a covering class, and that $s, t \in R$ are fixed with $0 < s < t$. Then $\xi \stackrel{d}{=} \eta$ iff η is a.s. simple (diffuse) and*

$$Ee^{-t\xi U} = Ee^{-t\eta U} , \qquad U \in \mathcal{U} , \tag{3.5}$$

and also iff (3.5) holds and in addition

$$Ee^{-s\xi C} \leqq Ee^{-s\eta C} , \qquad C \in \mathcal{C} . \tag{3.6}$$

If $E\xi \in \mathfrak{M}$, then (3.6) may be replaced by the condition

$$E\xi C \geqq E\eta C , \qquad C \in \mathcal{C} . \tag{3.7}$$

It is interesting to observe that (3.3) is formally obtained from (3.5) by letting $t \to \infty$.

Proof. Let us first consider the point process case, and let ξ_p and η_p be p-thinnings of ξ and η respectively. By Table 1,

$$L_{\xi_p}(f) = L_\xi(\log [1 - p (1 - e^{-f})]) , \qquad f \in \mathcal{F} ,$$

and putting $f = 1_B$, we get in particular

$$Ee^{-u\xi_p B} = E \exp \{\xi B \log [1 - p (1 - e^{-u})]\} , \quad B \in \mathcal{B}, \quad u \in R_+ . \tag{3.8}$$

As $u \to \infty$, we obtain by dominated convergence

$$P\{\xi_p B = 0\} = E \exp \{\xi B \log (1 - p)\} , \qquad B \in \mathcal{B} . \tag{3.9}$$

Choosing p such that $\log(1-p) = -t$, i.e. $p = 1 - e^{-t}$, it follows from (3.5), (3.9) and the corresponding relation for η and η_p that

$$P\{\xi_p U = 0\} = Ee^{-t\xi U} = Ee^{-t\eta U} = P\{\eta_p U = 0\}\,, \qquad U \in \mathscr{U}\,,$$

so Theorem 3.3 yields $\xi_p^* \stackrel{d}{=} \eta_p^*$. Now ξ and ξ_p are simultaneously a.s. simple, and correspondingly for η and η_p (cf. Exercise 2.4), so we get $\xi_p \stackrel{d}{=} \eta_p^*$ in general, and if η is a.s. simple, even $\xi_p \stackrel{d}{=} \eta_p$. In the latter case, Corollary 3.2 yields $\xi \stackrel{d}{=} \eta$, which proves the first assertion.

In proving the second assertion, we may assume that $\xi_p \stackrel{d}{=} \eta_p^*$ and that (3.6) holds. Since $0 < s < t$, there exists some $u > 0$ satisfying

$$1 - e^{-s} = (1 - e^{-t})\,(1 - e^{-u}) = p\,(1 - e^{-u})\,.$$

or equivalently

$$\log\,[1 - p(1 - e^{-u})] = -s\,.$$

Inserting this u into (3.8) and the corresponding relation for η, and using (3.6), we obtain for any $C \in \mathscr{C}$

$$Ee^{-u\eta_p^* C} = Ee^{-u\xi_p C} = Ee^{-s\xi C} \leqq Ee^{-s\eta C} = Ee^{-u\eta_p C}\,,$$

so

$$E\{e^{-u\eta_p^* C} - e^{-u\eta_p C}\} \leqq 0\,, \qquad C \in \mathscr{C}\,.$$

But here the random variable within brackets is non-negative, so it must in fact be a.s. zero, and we obtain $\eta_p^* C = \eta_p C$ a.s., $C \in \mathscr{C}$. Thus it follows as in the proof of Lemma 2.7 that η_p is a.s. simple, and the proof may be completed as before. The last assertion follows by a similar argument based on the relations $E\xi_p = pE\xi$ and $E\eta_p = pE\eta$ (cf. Table 1).

The random measure case may be proved from Theorem 3.3 by a similar argument, where instead of thinnings we consider Cox processes directed by $t\xi$ and $t\eta$. Alternatively, we may fix an arbitrary $u > t$ and apply the point process case of the present theorem to Cox processes directed by $u\xi$ and $u\eta$. The details are left to the reader. □

Compounds

Theorem 3.5. *Let ξ be a β-compound of some point process η on $\mathfrak{S}$. Further suppose that $p = P\{\beta > 0\} > 0$, and that $C \in \mathscr{B}$ is such that $C\eta$ is a.s. simple and $\eta C \stackrel{d}{\neq} 0$. Then $P\eta^{-1}$ and $P\beta^{-1}$ are uniquely determined by $P\xi^{-1}$, p and C.*

Proof. As in (3.8) we get from Table 1

$$Ee^{-t\xi B} = E\exp\,[\eta B \log L_\beta(t)]\,, \qquad B \in \mathscr{B}, \qquad t \in R_+\,, \tag{3.10}$$

and since $L_\beta(t) \to P\{\beta = 0\} = 1 - p$ as $t \to \infty$, it follows by dominated convergence that for arbitrary $B \in \mathscr{B}$

$$P\{\xi B = 0\} = \begin{cases} E\exp\,[\eta B \log(1-p)] & \text{if } 0 < p < 1\,, \\ P\{\eta B = 0\} & \text{if } p = 1\,. \end{cases}$$

Replacing B by $B \cap C$ and noting that $\eta(B \cap C) \equiv (C\eta)B$, it is seen from Theorem 3.4 or 3.3 respectively that $P(C\eta)^{-1}$ and hence also $P(\eta C)^{-1}$ is uniquely determined. Letting $B = C$ in (3.10), we further obtain

$$L_{\xi C} = L_{\eta C} \circ (-\log L_\beta) ,$$

and since $\eta C \overset{d}{\neq} 0$ by assumption, $L_{\eta C}$ has a unique inverse $L_{\eta C}^{-1}$ on the interval $(P\{\eta C = 0\}, 1]$, and we get

$$L_\beta = \exp\{- L_{\eta C}^{-1} \circ L_{\xi C}\} .$$

By A 5.1, this implies that even $P\beta^{-1}$ is unique, and so we may apply Corollary 3.2 to complete the proof. □

Notes. In Theorem 3.1, the implication (ii) ⇒ (i) (which is the only non-trivial one) is due to PROHOROV [69] (see also [28, 29, 76]). Theorem 3.3 was proved in [31], and independently (in a slightly weaker form) by MÖNCH [59], after the particular case for POISSON processes had been discovered by RÉNYI [71]. (See also KENDALL [40] for related results in a much more general setting.) As for Theorem 3.4, the point process case is new, while for random measures the first assertion is essentially proved by MÖNCH (cf. [44], p. 316), and independently in [35] and GRANDELL [18]. In Corollary 3.2, the thinning case is due to MECKE [57, 58] and the case of COX processes to KUMMER and MATTHES [47]; the results for general compounds, there and in Theorem 3.5, are new.

Exercises

3.1. Show that Theorem 3.1 remains true with $\mathcal{F}_c$ in (ii), replaced by the class of simple functions over $\mathcal{I}$. (Cf. Exercise 1.1 for a definition.)

3.2. Show that Theorem 3.3 (and hence also Theorem 3.4) remains true for any ring $\mathcal{U} \subset \mathcal{B}$ and semiring $\mathcal{I} \subset \mathcal{B}$ satisfying $\hat{\sigma}(\mathcal{U}) = \hat{\sigma}(\mathcal{I}) = \mathcal{B}$. (Hint: Use A 2.2 to extend (3.3) to $\mathcal{B}$, and then apply the result of Exercise 2.10. Cf. KERSTAN, MATTHES and MECKE [44], pp. 32, 43.)

3.3. Apply the method involving thinnings and Cox processes to the second assertion in Theorem 3.3 to obtain an alternative to condition (3.6) in Theorem 3.4.

3.4. Let $\mathcal{C} \subset \mathcal{B}$ be a covering class and let $s > 0$ be fixed. Show that Theorem 3.3 remains true with (3.4) replaced by (3.6) or (3.7).

3.5. Let ξ, η, $\mathcal{U}$ and t be such as in Theorem 3.4, and suppose that (3.5) holds. Show that, in the point process case

$$L_\eta(f) \leqq L_\xi(f) \leqq L_{\eta^*}(f) , \qquad f \in \mathcal{F} , \qquad \|f\| \leqq t ,$$

while in the random measure case,

$$L_\eta(f) \leqq L_\xi(f) \leqq L_{\eta_d}(f) , \qquad f \in \mathcal{F} , \qquad \|f\| \leqq t .$$

3.6. Prove that, if the class $\mathcal{C}$ in Theorem 3.4 is a DC-semiring, then ξ may be allowed to have atoms of fixed size and location. (Cf. the proof of Lemma 7.10 below.)

3.7. Assume that $\mathfrak{S}$ is countable. Show that there exists some fixed function $f \in \mathcal{F}$ such that $P\xi^{-1}$ is determined by $P(\xi f)^{-1}$ for any point process ξ on $\mathfrak{S}$ with $\xi\mathfrak{S} < \infty$ a.s. (Hint: Let the numbers $f(s)$, $s \in \mathfrak{S}$, be rationally independent and bounded above and below by positive constants.)

3.8. Show that the distribution of a point process ξ on $\mathfrak{S}$ is not, in general, determined by $P(\xi B)^{-1}$ for all $B \in \mathcal{B}$. (Cf. LEE [51], and [44], p. 17. Hint: It is enough to take $\mathfrak{S} = \{1, 2\}$ and assume that $\xi_3^* = 0$.)

Proof. Since by A 7.3 $D_{\pi_f} \subset \{\mu: \mu D_f > 0\}$, $f \in \mathcal{F}_c$, while

$$D_{\pi_{B_1,\ldots,B_k}} = \bigcup_{j=1}^{k} D_{\pi_{B_j}} \subset \bigcup_{j=1}^{k} \{\mu: \mu\partial B_j > 0\}\,, \qquad k \in N, \qquad B_1, \ldots, B_k \in \mathcal{B}\,,$$

the assertions are immidiate consequences of A 4.2. □

Lemma 4.5. *A sequence $\{\xi_n\}$ of random measures or point processes on $\mathfrak{S}$ is relatively compact w.r.t. convergence in distribution in the vague topology iff*

$$\lim_{t\to\infty} \limsup_{n\to\infty} P\{\xi_n B > t\} = 0\,, \qquad B \in \mathcal{B}\,. \tag{4.2}$$

Note that (4.2) is equivalent to tightness, and hence by A 4.4 to relative compactness of $\{\xi_n B\}$ for every $B \in \mathcal{B}$.

Proof. By A 4.4 and A 7.7, it is enough to prove that (4.2) is equivalent to tightness of $\{\xi_n\}$, and since $\mathfrak{N}$ is closed in $\mathfrak{M}$ (cf. A 7.4), we need only consider the random measure case. The proof is based on the fact (cf. A 7.5) that a set $M \subset \mathfrak{M}$ is relatively compact iff

$$\sup_{\mu\in M} \mu B < \infty, \qquad B \in \mathcal{B}\,. \tag{4.3}$$

Let us first suppose that $\{\xi_n\}$ is tight. This means that there exists for every $\varepsilon > 0$ some compact set $M \subset \mathfrak{M}$ satisfying $P\{\xi_n \notin M\} < \varepsilon$, $n \in N$, and hence

$$P\Big\{\xi_n B > \sup_{\mu\in M} \mu B\Big\} \leqq P\{\xi_n \notin M\} < \varepsilon\,, \qquad n \in N, \qquad B \in \mathcal{B}\,.$$

Since (4.3) holds for this M, we obtain (4.2).

Suppose conversely that (4.2) holds. Choose a sequence $G_1, G_2, \ldots$ of open $\mathcal{B}$-sets satisfying $G_n \uparrow \mathfrak{S}$ (cf. A 6.1), and note that there exist for every fixed $\varepsilon > 0$ some constants $c_1, c_2, \ldots \in R_+$ with

$$P\{\xi_n G_k > c_k\} \leqq \varepsilon 2^{-k-1}\,, \qquad k, n \in N\,.$$

The set

$$M = \bigcap_{k=1}^{\infty} \{\mu \in \mathfrak{M}:\ \mu G_k \leqq c_k\}$$

is relatively compact, since every fixed $B \in \mathcal{B}$ is contained in some G_k, and hence

$$\sup_{\mu\in M} \mu B \leqq \sup_{\mu\in M} \mu G_k \leqq c_k < \infty\,,$$

proving (4.3). The tightness of $\{\xi_n\}$ now follows from the fact that

$$P\{\xi_n \notin M^-\} \leqq P\{\xi_n \notin M\} = P \bigcup_{k=1}^{\infty} \{\xi_n G_k > c_k\} \leqq \sum_{k=1}^{\infty} P\{\xi_n G_k > c_k\}$$

$$\leqq \varepsilon \sum_{k=1}^{\infty} 2^{-k-1} = \varepsilon\,.$$ □

Lemma 4.6. *Let $\xi, \xi_1, \xi_2, \ldots$ be random measures on $\mathfrak{S}$ and let $\mathcal{U} \subset \mathcal{B}$ be a DC-ring. Further suppose that*

$$\liminf_{n\to\infty} Ee^{-t\xi_n U} \geqq Ee^{-t\xi U}\,, \qquad U \in \mathcal{U}\,, \tag{4.4}$$

for some fixed $t > 0$. Then $\mathcal{B}_\eta \supset \mathcal{B}_\xi$ for any random measure η such that $\xi_n \xrightarrow{d} \eta$ as $n \to \infty$ through some subsequence. In the point process case, the assertion *remains true for* $t = \infty$, *i.e. with* (4.4) *replaced by*

$$\liminf_{n\to\infty} P\{\xi_n U = 0\} \geqq P\{\xi U = 0\} , \qquad U \in \mathcal{U} . \tag{4.5}$$

Proof. Let $B \in \mathcal{B}_\xi$ and $\varepsilon < 0$ be arbitrary. By Lemma 4.3 we may choose some closed set $C \in \mathcal{B}_\eta$ with $C \supset \partial B$ and such that $Ee^{-t\xi(C\setminus\partial B)} \geqq 1 - \varepsilon$. We may next choose some $U \in \mathcal{U}$ with $U \supset C$ such that $Ee^{-t\xi(U\setminus C)} \geqq 1 - \varepsilon$. Since the function $1 - e^{-x}$ is subadditive on R_+, we then obtain

$$1 - Ee^{-t\xi U} \leqq (1 - Ee^{-t\xi\partial B}) + (1 - Ee^{-t\xi(C\setminus\partial B)}) + (1 - Ee^{-t\xi(U\setminus C)}) \leqq 2\varepsilon .$$

If $\xi_n \xrightarrow{d} \eta$ $(n \in N')$, we may thus conclude from (4.4) and Lemma 4.4 that

$$Ee^{-t\eta\partial B} \geqq Ee^{-t\eta C} = \lim_{n\in N'} Ee^{-t\xi_n C} \geqq \liminf_{n\to\infty} Ee^{-t\xi_n C}$$
$$\geqq \liminf_{n\to\infty} Ee^{-t\xi_n U} \geqq Ee^{-t\xi U} \geqq 1 - 2\varepsilon ,$$

and since ε was arbitrary, it follows that $\eta\partial B = 0$ a.s., i.e. that $B \in \mathcal{B}_\eta$. The last assertion may be proved by a similar argument. □

Proof of Theorem 4.2. Since (i) implies (ii) and (iii) by Lemma 4.4, while (ii) implies (ii)′ by A 4.2, it is enough to prove that (ii)′ and (iii) both imply (i). Let us first suppose that (ii)′ holds. Then

$$L_{\xi_n f}(t) = L_{\xi_n}(tf) \to L_\xi(tf) = L_{\xi f}(t) , \qquad f \in \mathcal{F}_c, \quad t \in R_+ ,$$

and so (ii) holds by A 5.2. By A 4.4, A 6.1 and Lemma 4.5, $\{\xi_n\}$ is then relatively compact, so any sequence $N' \subset N$ must contain some subsequence N'' such that $\xi_n \xrightarrow{d}$ some η $(n \in N'')$. By Lemma 4.4 we obtain $\xi_n f \xrightarrow{d} \eta f$ $(n \in N'')$, $f \in \mathcal{F}_c$, and comparing with (ii), it is seen that $\xi f \overset{d}{=} \eta f$, $f \in \mathcal{F}_c$. But by Theorem 3.1, this implies $\xi \overset{d}{=} \eta$, so we have in fact $\xi_n \xrightarrow{d} \xi$ $(n \in N'')$. Thus (i) follows by A 1.2.

Let us next suppose that (iii) holds. Then we may argue as above, except that it is not obvious that $\xi_n \xrightarrow{d} \eta$ $(n \in N'')$ implies

$$(\xi_n I_1, \ldots , \xi_n I_k) \xrightarrow{d} (\eta I_1, \ldots , \eta I_k) \ (n \in N''), \ k \in N, \ I_1, \ldots , I_k \in \mathcal{I} . \tag{4.6}$$

For a justification, let $\mathcal{U}$ be the ring generated by $\mathcal{I}$ and note that (iii) implies $\xi_n U \xrightarrow{d} \xi U, U \in \mathcal{U}$, by A 4.2. Hence (4.4) holds for any $t > 0$, and we get $\mathcal{B}_\eta \supset \mathcal{B}_\xi$ by Lemma 4.6 whenever $\xi_n \xrightarrow{d} \eta$ $(n \in N'')$. In particular $\mathcal{I} \subset \mathcal{B}_\eta$, and so (4.6) follows by Lemma 4.4. □

Simplicity and diffuseness

We are now going to establish partial improvements of Theorem 4.2. Given any covering class $\mathcal{I} \subset \mathcal{B}$ and any $a > 0$, we shall say that a sequence $\{\xi_n\}$ of random measures on $\mathfrak{S}$ is *a-regular* w.r.t. $\mathcal{I}$ if for every $I \subset \mathcal{I}$ there exists some array $\{I_{mk}\} \subset \mathcal{I}$ of finite covers of I (one for each $m \in N$) such that

$$\lim_{m\to\infty} \limsup_{n\to\infty} \sum_k P\{\xi_n I_{mk} \geqq a\} = 0 . \tag{4.7}$$

5. Existence

The uniqueness results of Section 3 give rise to a number of existence problems, some of which will be discussed in this section. Our approach will be based on the following extensions of results in Section 4.

Limit unknown

Lemma 5.1. *Let $\xi_1, \xi_2, \ldots$ be random measures on $\mathfrak{S}$ and let $\mathcal{U} \subset \mathcal{B}$ be a DC-ring. Suppose that either*

(i) $\xi_n f \xrightarrow{d}$ some ξ_f, $f \in \mathcal{F}_c$, *or*

(ii) $(\xi_n U_1, \ldots, \xi_n U_k) \xrightarrow{d}$ *some* $\xi_{U_1,\ldots,U_k}$, $k \in N$, $U_1, \ldots, U_k \in \mathcal{U}$, *where the* ξ_U *are such that* $\xi_{U_m} \xrightarrow{d} 0$ *wheneve* $U, U_1, U_2, \ldots \in \mathcal{U}$ *with* $U_m \downarrow \partial U$.

Then $\xi_n \xrightarrow{d}$ *some* ξ *satisfying* $\xi f \stackrel{d}{=} \xi_f$, $f \in \mathcal{F}_c$, *or* $(\xi U_1, \ldots, \xi U_k) \stackrel{d}{=} \xi_{U_1,\ldots,U_k}$, $k \in N$, $U_1, \ldots, U_k \in \mathcal{U}$, *respectively.*

Proof. In either case, $\{\xi_n\}$ is relatively compact by Lemma 4.5, so for some subsequence $N' \subset N$, $\xi_n \xrightarrow{d}$ some ξ $(n \in N')$. But then $\xi_n f \xrightarrow{d} \xi f$ $(n \in N')$, $f \in \mathcal{F}_c$, by Lemma 4.4, so assuming (i), we obtain $\xi f \stackrel{d}{=} \xi_f$, $f \in \mathcal{F}_c$, and the asserted convergence follows by Theorem 4.2.

In case of (ii), the same argument goes through, provided we can show that $\mathcal{U} \subset \mathcal{B}_\xi$. For this purpose, let $U \in \mathcal{U}$ and $\varepsilon > 0$ be arbitrary, and note that $(\partial U)_\delta = \{s \in \mathfrak{S}\colon \varrho(s, \partial U) \leqq \delta\}$ is bounded for sufficiently small $\delta > 0$. For every such δ we may choose $C \in \mathcal{B}_\xi$ and $V \in \mathcal{U}$ such that $\partial U \subset C \subset (\partial U)_\delta \subset V \subset (\partial U)_{2\delta}$, and here we may take δ so small that $P\{\xi_V \geqq \varepsilon\} \leqq \varepsilon$. Then

$$P\{\xi \partial U > \varepsilon\} \leqq P\{\xi C > \varepsilon\} \leqq \liminf_{n \in N'} P\{\xi_n C > \varepsilon\} \leqq \limsup_{n \to \infty} P\{\xi_n V \geqq \varepsilon\}$$

$$\leqq P\{\xi_V \geqq \varepsilon\} \leqq \varepsilon \,,$$

and since ε was arbitrary, we get $U \in \mathcal{B}_\xi$, as desired. □

The proof of the next result is similar and may be left to the reader.

Lemma 5.2. *Let $\xi_1, \xi_2, \ldots$ be point processes (or random measures) on $\mathfrak{S}$, and suppose that $\mathcal{U} \subset \mathcal{B}$ is a DC-ring while $\mathcal{I} \subset \mathcal{B}$ is a DC-semiring. Further suppose that $\{\xi_n\}$ is 2-regular (or regular respectively) w.r.t. $\mathcal{I}$ and that, for some fixed $t > 0$,*

$$Ee^{-t\xi_n U} \to \textit{some } \varphi_t(U), \qquad U \in \mathcal{U} \,.$$

Finally suppose that $\varphi_t(U_m) \to 1$ whenever $U_1, U_2, \ldots \in \mathcal{U}$ and $V \in \mathcal{U} \cup \mathcal{I}$ with $U_m \downarrow \partial V$. Then there exists some a.s. simple point process (or diffuse random measure) ξ on $\mathfrak{S}$ such that $\xi_n \xrightarrow{d} \xi$ and $\varphi_t(U) = Ee^{-t\xi U}$, $U \in \mathcal{U}$. In the point process case, we may even take $t = \infty$.

General existence criteria

Let ξ be a random measure on $\mathfrak{S}$ and define $P_{B_1,\dots,B_k} = P(\xi B_1, \dots, \xi B_k)^{-1}$, $k \in N$, $B_1, \dots, B_k \in \mathcal{B}$. Then the projection of $P_{B_1,\dots,B_k}$ onto the subspace spanned by the coordinate set $\{j_1, \dots, j_h\} \subset \{1, \dots, k\}$ coincides with $P_{B_{j_1},\dots,B_{j_h}}$. We shall say that a family $\{P_{B_1,\dots,B_k}\}$ of probability measures is *consistent* if it satiesfies all relations of this type. In that case there exists by KOLMOGOROV's consistency theorem some random process ξ on $\mathcal{B}$ satisfying

$$P\big(\xi(B_1), \dots, \xi(B_k)\big)^{-1} = P_{B_1,\dots,B_k}, \quad k \in N, \ B_1, \dots, B_k \in \mathcal{B} .$$

Additional conditions are needed to ensure ξ to have a measure valued version.

Theorem 5.3. *Let $\mathcal{U} \subset \mathcal{B}$ be a DC-ring and let $P_{U_1,\dots,U_k}$, $k \in N$, $U_1, \dots, U_k \in \mathcal{U}$, be a consistent family of probability measures. Then there exists some random measure ξ on $\mathfrak{S}$ with $\mathcal{U} \subset \mathcal{B}_\xi$ and $P(\xi U_1, \dots, \xi U_k)^{-1} = P_{U_1,\dots,U_k}$, $k \in N$, $U_1, \dots, U_k \in \mathcal{U}$, iff*

(i) $P_{U,V,U\cup V}\{(x, y, z)\colon x + y = z\} = 1$ *for all disjoint* $U, V \in U$,

(ii) $P_{U_n} \xrightarrow{w} \delta_0$ *whenever* $U, U_1, U_2, \dots \in \mathcal{U}$ *with* $U_n \downarrow \partial U$.

Here $\xrightarrow{w}$ denotes weak convergence.

Proof. If ξ is a random measure on $\mathfrak{S}$ with the stated properties, then (i) and (ii) follow from the facts that $\xi U + \xi V = \xi(U \cup V)$ for disjoint U and V and that $\xi U_n \downarrow \xi \partial U = 0$ a.s. whenever $U_n \downarrow \partial U$, $U \in \mathcal{B}_\xi$.

Suppose conversely that (i) and (ii) hold. For fixed $U \in \mathcal{U}$, let $\{U_{nj}\} \subset \mathcal{U}$ be a null-array of nested partitions of U and choose for each $n \in N$ some random measure ξ_n on $\mathfrak{S}$ with all its mass confined to U and such that

$$P(\xi_n U_{n1}, \ \xi_n U_{n2}, \dots)^{-1} = P_{U_{n1}, U_{n2}, \dots} .$$

(For the existence, note that we can choose a fixed point s_{nj} in each U_{nj} and let the mass in U_{nj} be confined to s_{nj}.) Writing $\mathcal{U}_n$ for the ring generated by $U_{n1}, U_{n2}, \dots$, it follows by (i) that

$$P(\xi_n U_1, \dots, \xi_n U_k)^{-1} = P_{U_1,\dots,U_k}, \quad k, n \in N, \ U_1, \dots, U_k \in \mathcal{U}_m, \ m \leqq n ,$$

and hence

$$P(\xi_n U_1, \dots, \xi_n U_k)^{-1} \xrightarrow{w} P_{U_1,\dots,U_k}, \quad k \in N, \ U_1, \dots, U_k \in \mathcal{U}' ,$$

where $\mathcal{U}' = \bigcup_m \mathcal{U}_m$. Since $\mathcal{U}'$ can easily be extended to a DC-ring on the whole space $\mathfrak{S}$, it follows by (ii) and Lemma 5.1 that $\xi_n \xrightarrow{d}$ some ξ' satisfying $\xi' U^c = 0$ a.s. and

$$P(\xi' U_1, \dots, \xi' U_k)^{-1} = P_{U_1,\dots,U_k}, \quad k \in N, \ U_1, \dots, U_k \in \mathcal{U}' . \tag{5.1}$$

To see that (5.1) extends to $\mathcal{U} \cap U$, let $V_1, \dots, V_m \in \mathcal{U} \cap U$ be arbitrary and replace $\mathcal{U}'$ in the above argument by a refined null-array $\mathcal{U}''$ containing $V_1, \dots, V_m$. We then obtain

$$P(\xi'' U_1, \dots, \xi'' U_k)^{-1} = P_{U_1,\dots,U_k}, \quad k \in N, \ U_1, \dots, U_k \in \mathcal{U}'' , \tag{5.2}$$

for some ξ'', and comparing with (5.1). it follows by Theorem 3.1 that $\xi' \stackrel{d}{=} \xi''$. Inserting this into (5.2) shows that (5.1) remains true for $V_1, \dots, V_m$, which is the desired extension of (5.1).

To complete the proof, let $U_1, U_2, \dots \in \mathcal{U}$ with $U_n^\circ \uparrow \mathfrak{S}$, ($U_n^\circ$ being the interior of U_n), and let $\xi_1', \xi_2', \dots$ be random measures constructed as ξ' above but with U replaced by $U_1, U_2, \dots$ Then it follows by another application of Lemma 5.1 that $\xi_n' \xrightarrow{d}$ some ξ with the stated properties. □

Theorem 5.4. *Let $P_{B_1, \dots, B_k}$, $k \in N$, $B_1, \dots, B_k \in \mathcal{B}$, be a consistent family of probability measures. Then there exists some random measure ξ on $\mathfrak{S}$ satisfying $P(\xi B_1, \dots, \xi B_k)^{-1} = P_{B_1, \dots, B_k}$, $k \in N$, $B_1, \dots, B_k \in \mathcal{B}$, iff*

(i) $P_{B, C, B \cup C}\{(x, y, z): x + y = z\} = 1$ *for all disjoint* $B, C \in \mathcal{B}$,

(ii) $P_{B_n} \xrightarrow{w} \delta_0$ *whenever* $B_1, B_2, \dots \in \mathcal{B}$ *with* $B_n \downarrow \emptyset$.

As usual, $\emptyset$ denotes the empty set.

Proof. The necessity is proved as before. Suppose conversely that (i) and (ii) hold, and define $\mathcal{U} = \{B \in \mathcal{B}: P_{\partial B} = \delta_0\}$. Then it is seen from (i) and A 1.1 that $\mathcal{U}$ is a ring, and we shall show that it is even a DC-ring. To see this, it suffices as in case of Lemma 4.3 to prove that, for any fixed $t \in \mathfrak{S}$, the ball $S_r = \{s \in \mathfrak{S}: \varrho(s, t) \leqq r\}$ belongs to $\mathcal{U}$ for arbitrarily small $r > 0$. If this statement were false, there would exist some $\varepsilon > 0$ and some sequence $r_1 > r_2 > \cdots > 0$ such that $S_{r_1} \in \mathcal{B}$ and $P\{\xi \partial S_{r_n} \geqq \varepsilon\} \geqq \varepsilon$, $n \in N$, where ξ is a random process on $\mathcal{B}$ with marginal distributions $P_{B_1, \dots, B_k}$. Since by (i) $\sum_n \xi \partial S_{r_n} \leqq \xi S_{r_1} < \infty$ a.s., Fatou's lemma would then yield the contradiction

$$0 = P\left\{\sum_n \xi \partial S_{r_n} = \infty\right\} \geqq P \limsup_{n\to\infty} \{\xi \partial S_{r_n} \geqq \varepsilon\} \geqq \limsup_{n\to\infty} P\{\xi \partial S_{r_n} \geqq \varepsilon\} \geqq \varepsilon .$$

We may now apply Theorem 5.3 to conclude that there exists some random measure ξ on $\mathfrak{S}$ satisfying

$$P(\xi U_1, \dots, \xi U_k)^{-1} = P_{U_1, \dots, U_k}, \quad k \in N, \quad U_1, \dots, U_k \in \mathcal{U} . \tag{5.3}$$

Indeed, condition (ii) of that theorem is fulfilled, since clearly $P_{B_n} \xrightarrow{w} \delta_0$ implies $P_{B, B_n} \xrightarrow{w} P_B \times \delta_0$ for every $B \in \mathcal{B}$. By the same argument, it is seen from (i) and (ii) that $P_{B_1, \dots, B_k}$ is continuous under bounded monotone convergence of each set $B_1, \dots, B_k$, and since this is also true for $P(\xi B_1, \dots, \xi B_k)^{-1}$, we may apply A 2.2 to one component at a time to extend (5.3) to $\mathcal{B}$. □

The point process case

As a prelude to the point process case, we consider a general necessary condition. Given any function $\varphi: \mathcal{B} \to R$ and any $A_1, \dots, A_n \in \mathcal{B}$, $n \in N$, we define the set function $\Delta_{A_1} \cdots \Delta_{A_n}\varphi$ by putting

$$\Delta_A \varphi(B) = \varphi(A \cup B) - \varphi(B) , \qquad A, B \in \mathcal{B} ,$$

and then proceeding recursively. We shall say that φ is *completely monotone* if

$$(-1)^n \Delta_{A_1} \cdots \Delta_{A_n}\varphi(B) \geqq 0 , \qquad n \in N, \qquad A_1, \dots, A_n, B \in \mathcal{B} .$$

Lemma 5.5. *Let ξ be a random measure on $\mathfrak{S}$ and let $0 < t \leqq \infty$. Then the set function $\varphi_t(B) = Ee^{-t\xi B}$, $B \in \mathscr{B}$, is completely monotone.*

Proof. For any $A, B \in \mathscr{B}$ we get

$$-\Delta_A P\{\xi B = 0\} = P\{\xi B = 0\} - P\{\xi(A \cup B) = 0\}$$
$$= P\{\xi B = 0,\ \xi(A \cup B) > 0\} = P\{\xi B = 0,\ \xi A > 0\} \geqq 0\,,$$

and by induction it is seen that more generally

$$(-1)^n \Delta_{A_1} \cdots \Delta_{A_n} P\{\xi B = 0\} = P\{\xi B = 0,\ \xi A_1 > 0, \ldots, \xi A_n > 0\} \geqq 0$$

for arbitrary $n \in N$ and $A_1, \ldots, A_n, B \in \mathscr{B}$. This proves the assertion for $t = \infty$. For $t < \infty$ we may reduce the proof to this case by considering a Cox process η directed by $t\xi$ and noting that

$$Ee^{-t\xi B} = P\{\eta B = 0\}, \qquad B \in \mathscr{B}\,. \qquad \square$$

Let us now suppose conversely that φ is a completely monotone set function defined on some DC-ring $\mathscr{U} \subset \mathscr{B}$ and such that $\varphi(\emptyset) = 1$. Let $V \in \mathscr{U}$ be fixed, and consider a null-array $U_{nj} \in \mathscr{U}$, $j = 1, \ldots, k_n$, $n \in N$, of nested partitions of V. For each $n \in N$, it is then possible to define a point process ξ_n on $\mathfrak{S}$ with all its mass confined to V, such that $\xi_n U_{nj}$ equals 0 or 1 for every j and moreover

$$P\{\xi_n U_{nj} = 1,\ j \in J;\ \xi_n U_{nj} = 0,\ j \notin J\} = \Big\{\prod_{j \in J} (-\Delta_{U_{nj}})\Big\}\, \varphi\big(\bigcup_{j \notin J} U_{nj}\big) \tag{5.4}$$

for every subset $J \subset \{1, \ldots, k_n\}$. In fact, it follows from the complete monotonicity of φ that these quantities are non-negative. Furthermore, summation over all J yields 1, as may easily be verified by induction in the number of partitioning sets. Finally, it is seen that

$$P\{\xi_n U = 0\} = \varphi(U)\,, \qquad U \in \mathscr{U}_m, \quad m, n \in N, \quad m \leqq n\,,$$

where $\mathscr{U}_m$ is the ring generated by $U_{m1}, \ldots, U_{mk_m}$, so we get

$$P\{\xi_n U = 0\} \to \varphi(U), \quad \mathscr{U} \in \bigcup_m \mathscr{U}_m\,.$$

Let us now assume that $\varphi(U_m) \to 1$ whenever $U, U_1, U_2, \ldots \in \mathscr{U}$ with $U_m \downarrow \partial U$, and further that the sequence $\{\xi_n\}$ constructed above is relatively compact. Proceeding as in the proof of Theorem 5.3, we may then conclude that there exists some point process ξ on $\mathfrak{S}$ satisfying

$$\varphi(U) = P(\xi U = 0\}\,, \qquad U \in \mathscr{U}\,. \tag{5.5}$$

Since ξ^* satisfies the same relation, we have thus proved

Theorem 5.6. *Let $\mathscr{U} \subset \mathscr{B}$ be a DC-ring and let φ be a completely monotone function on $\mathscr{U}$ such that $\varphi(\emptyset) = 1$. Then there exists some a.s. simple point process ξ on $\mathfrak{S}$ satisfying (5.5) iff*

(i) *$\varphi(U_n) \to 1$ whenever $U, U_1, U_2, \ldots \in \mathscr{U}$ with $U_n \downarrow \partial U$,*
(ii) *for every $U \in \mathscr{U}$ there exists some null-array $\{U_{nj}\} \subset \mathscr{U}$ of nested partitions of U such that, for ξ_n as in (5.4), $n \in N$, the sequence $\{\xi_n U\}$ is tight.*

The last condition may easily be expressed directly in terms of φ, but the present form seems preferable for applications.

Notes. Theorem 5.4 is due with an entirely different proof to NAWROTZKI [62] and HARRIS [21, 22, 23] (cf. [28, 29, 27]). An extended version of this result is given (for point processes) by KERSTAN, MATTHES and MECKE [44], p. 17. See also PROHOROV [69] and MECKE [58] for different types of existence criteria. Theorem 5.6 is essentially a restatement of a theorem by KARBE (cf. [44], p. 35) and KURTZ [48]. We finally refer to JIŘINA [29], p. 9, and MECKE [58], p. 13, for results related to Lemma 5.1.

Exercises

5.1. Let $\xi_1, \xi_2, \ldots$ be random measures on $\mathfrak{S}$ such that $L_{\xi_n}(f) \to$ some c_f, $f \in \mathcal{F}_c$, where $c_f \to 1$ as $f \downarrow 0$. Show that there exists some random measure ξ on $\mathfrak{S}$ with $L_\xi(f) \equiv c_f$ and $\xi_n \xrightarrow{d} \xi$. (Hint: Apply A5.2 to reduce to Lemma 5.1.)

5.2. Let $P_{B_1, \ldots, B_k}$, $k \in N$, $B_1, \ldots, B_k \in \mathcal{B}$, be a consistent family of probability measures, and let ξ be a random process on $\mathcal{B}$ with these measures as finite-dimensional distributions. Show that condition (i) of Theorem 5.4 implies that $\xi \bigcup_{j=1}^{k} B_k = \sum_{j=1}^{k} \xi B_j$ a.s. for every $k \in N$ and disjoint $B_1, \ldots, B_k \in \mathcal{B}$. Show also by a direct argument that if (ii) of Theorem 5.4 holds in addition, then it is possible to choose a fixed exceptional null-set, i.e. ξ is a.s. finitely additive. (Hint: Approximate by sets in a fixed countable DC-ring.)

5.3. Suppose that $\mathfrak{S}$ is discrete, i.e. that all its points are isolated. Show that Theorem 5.6 then remains true even without conditions (i) and (ii).

5.4. Show that Theorem 5.6 is false in general without condition (i). (Hint: Let $\mathfrak{S} = R$ and place an atom at the fictitious point 0^+. Let $\mathcal{U}$ be the ring generated by all intervals $(a, b]$.)

5.5. Show that Theorem 5.6 is false in general without condition (ii). (Cf. [44], p. 42. Hint: Consider a "point process" on R with exactly one limit point α, where α is a random variable with diffuse distribution.)

5.6. Prove Lemma 5.5 for $t < \infty$ by a direct argument. (Hint: Let $\Delta_h f(t) = f(t + h) - f(t)$. Show that $(-1)^n \Delta_{h_1} \cdots \Delta_{h_n} e^{-t} \geqq 0$ by writing this expression as an n-fold integral of e^{-x}.)

5.7. Let $\mathcal{U} \subset \mathcal{B}$ be a countable DC-ring. Show that if condition (ii) of Theorem 5.6 is suppressed, then the set function φ there will still define the (unique) distribution of a random element ξ in $(\mathfrak{M}_\mathcal{U}, \mathcal{M}_\mathcal{U})$, as defined in Exercise 1.8. (Hint: Use KOLMOGOROV's consistency theorem to define a family ξ_U, $U \in \mathcal{U}$, of 0-1-valued random variables satisfying

$$P\{\xi_{U_1} = \cdots = \xi_{U_h} = 1;\ \xi_{V_1} = \cdots = \xi_{V_k} = 0\} = (-1)^h \Delta_{U_1} \cdots \Delta_{U_h} \varphi\left(\bigcup_{j=1}^{k} V_j\right)$$

for arbitrary $h, k \in Z_+$ and $U_1, \ldots, U_h, V_1, \ldots, V_k \in \mathcal{U}$. Then consider for fixed $U \in \mathcal{U}$ a null-array $\Pi_n = \{U_{nj}\} \subset \mathcal{U}$, $n \in N$, of nested partitions of U such that $\{U_{nj}\}$ generates $\mathcal{U} \cap U$, and define

$$\xi = \sum \left\{\delta_s\colon s = \bigcap_{n=1}^{\infty} U_n^-, U_n \in \Pi_n, U_n \supset U_{n+1}, \xi_{U_n} = 1, n \in N\right\}.$$

Then (i) yields $\xi\, \partial U = 0$ a.s., $U \in \mathcal{U}$, so we can replace U_n^- above by U_n. This result is new but the approach is due to KURTZ [48].)

6. Infinite divisibility, generalities

Here and in Section 7, our main purpose is to discuss the convergence in distribution of sums $\sum_j \xi_{nj}$ of random measures, where the ξ_{nj}, j, $n \in N$, form a *null-array* in the sense that they are independent for fixed n and that moreover

$$\lim_{n\to\infty} \sup_j P\{\xi_{nj}B > \varepsilon\} = 0\,, \qquad \varepsilon > 0\,, \qquad B \in \mathscr{B}\,. \tag{6.1}$$

Note that this condition reduces in case of point processes to

$$\lim_{n\to\infty} \sup_j P\{\xi_{nj}B > 0\} = 0\,, \qquad B \in \mathscr{B}\,.$$

It may be worth stressing that, for fixed n, the number of non-zero ξ_{nj} is allowed to be infinite.

Just as for random variables, the infinitely divisible distributions constitute the class of possible limit laws. A random measure ξ is said to be *infinitely divisible* (w.r.t. addition), if for each $n \in N$ there exist some independent and identically distributed random measures $\xi_1, \ldots, \xi_n$ such that $\xi \overset{d}{=} \xi_1 + \cdots + \xi_n$. In the point process case we require $\xi_1, \ldots, \xi_n$ to be point processes if nothing else is stated. Note that there exist point processes ξ which are infinitely divisible as random measures but not as point processes. (For a trivial example, let ξ be non-random and $\neq 0$.)

Representation and convergence

From here on we reserve the letter $\mathfrak{g}$ to denote the function $1 - e^{-x}$. A formula containing the symbol "limsi" is supposed to hold with both limsup and liminf at the actual place. For any measure λ on $\mathfrak{M}$ we define $\mathscr{B}_\lambda = \{B \in \mathscr{B}: \lambda\{\mu\partial B > 0\} = 0\}$. Write $\xrightarrow{v}$ for vague convergence.

Theorem 6.1. *The relation*

$$-\log Ee^{-\xi f} = \alpha f + \lambda(1 - e^{-\pi f})\,, \qquad f \in \mathscr{F}\,, \tag{6.2}$$

defines a unique correspondence between the distributions of all infinitely divisible random measures ξ on $\mathfrak{S}$ and the class of all pairs (α, λ), where $\alpha \in \mathfrak{M}$ while λ is a measure on $\mathfrak{M}\setminus\{0\}$ satisfying $(\lambda\pi_B^{-1})\,\mathfrak{g} < \infty$, $B \in \mathscr{B}$. If $\{\xi_{nj}\}$ is a null-array of random measures on $\mathfrak{S}$, then $\sum_j \xi_{nj} \xrightarrow{d}$ some ξ iff

$$\sum_j (1 - Ee^{-\xi_{nj}f}) \to \alpha f + \lambda(1 - e^{-\pi f})\,, \qquad f \in \mathscr{F}_c\,, \tag{6.3}$$

for some α and λ as above, and in this case $P\xi^{-1}$ is given by (6.2). *For any DC-semiring $\mathscr{I} \subset \mathscr{B}_\alpha \cap \mathscr{B}_\lambda$,* (6.3) *is equivalent to the conditions*

(i) $\sum_j P(\xi_{nj}I_1, \ldots, \xi_{nj}I_k)^{-1} \xrightarrow{v} \lambda\pi^{-1}_{I_1,\ldots,I_k}$ *in* $\mathfrak{M}(R_+^k\setminus\{0\})\,, \quad k \in N\,,$
$I_1, \ldots, I_k \in \mathscr{I}\,,$

(ii) $\lim_{\varepsilon\to 0} \operatorname{limsi}_{n\to\infty} \sum_j E[\xi_{nj}I\,; \xi_{nj}I < \varepsilon] = \alpha I\,, \quad I \in \mathscr{I}\,,$

(iii) $\lim_{r\to\infty} \operatorname{limsup}_{n\to\infty} \sum_j P\{\xi_{nj}I > r\} = 0\,, \quad I \in \mathscr{I}\,.$

In the point process case, we have $\alpha = 0$ while λ is confined to $\mathfrak{N}\setminus\{0\}$, and (i)—(iii) *are equivalent to*

$$\sum_j P(\xi_{nj}I_1, \ldots, \xi_{nj}I_k)^{-1} \xrightarrow{w} \lambda\pi^{-1}_{I_1,\ldots,I_k} \text{ in } \mathfrak{M}(Z_+^k\setminus\{0\})\,, \qquad k \in N\,, \qquad I_1, \ldots, I_k \in \mathscr{I}\,.$$

In proving this result, we shall proceed in three steps.

Proof in the one-dimensional case. If $\mathfrak{S}$ consists of exactly one point, then a random measure ξ on $\mathfrak{S}$ is completely determined by the random variable $\xi\mathfrak{S}$, and so we may identify these two quantities. In doing so, (6.1) may be written

$$\lim_{n\to\infty} \sup_j P\{\xi_{nj} > \varepsilon\} = 0 \,, \qquad \varepsilon > 0 \,,$$

and from A 5.2 it is seen that this is in turn equivalent to

$$\lim_{n\to\infty} \sup_j \big(1 - \varphi_{nj}(t)\big) = 0 \,, \qquad t \geqq 0 \,, \tag{6.4}$$

where $\varphi_{nj} = L_{\xi_{nj}}$. Let us now suppose that $\sum_j \xi_{nj} \xrightarrow{d}$ some ξ. Then A 5.2 yields $\prod_j \varphi_{nj} \to \varphi = L_\xi$, so $-\sum_j \log \varphi_{nj} \to \psi = -\log \varphi$. By (6.4), we may replace $\log \varphi_{nj}$ here by the first term in its Taylor expansion, thus obtaining

$$\psi_n \equiv \sum_j (1 - \varphi_{nj}) \to \psi \,. \tag{6.5}$$

Now

$$\psi_n(t+1) - \psi_n(t) = \sum_j \big(\varphi_{nj}(t) - \varphi_{nj}(t+1)\big) = \sum_j \int (e^{-tx} - e^{-(t+1)x})\, P\xi_{nj}^{-1}(dx)$$
$$= \int e^{-tx} \sum_j \mathfrak{g} P\xi_{nj}^{-1}(dx) \,,$$

and since ψ is continuous, it follows by (6.5) and A 5.2 that $\sum_j \mathfrak{g}P\xi_{nj}^{-1}$ is weakly convergent, i.e. that

$$\sum_j \mathfrak{g}P\xi_{nj}^{-1} \xrightarrow{w} \alpha\delta_0 + \mathfrak{g}\lambda \,, \tag{6.6}$$

for some $\alpha \in R_+$ and some $\lambda \in \mathfrak{M}(R'_+)$ with $\lambda\mathfrak{g} < \infty$. This proves (i)—(iii), while (6.3) may be deduced from (6.6) as follows.

$$\psi_n(t) = \sum_j (1 - Ee^{-t\xi_{nj}}) = \sum_j \int (1 - e^{-tx})\, P\xi_{nj}^{-1}(dx) = \int (1 - e^{-tx}) \sum_j P\xi_{nj}^{-1}(dx)$$
$$= \int \frac{1 - e^{-tx}}{1 - e^{-x}} \sum_j \mathfrak{g}P\xi_{nj}^{-1}(dx) \to \alpha t + \int \frac{1 - e^{-tx}}{1 - e^{-x}}\, \mathfrak{g}\lambda(dx)$$
$$= \alpha t + \int (1 - e^{-tx})\, \lambda(dx) = \psi(t) \,.$$

Conversely, it may be seen as in A 7.6 that (i)—(iii) imply (6.6), and hence that (6.5) holds for some continuous ψ. Reversing the above arguments, it then follows that $\prod_j \varphi_{nj} \to e^{-\psi}$, and the limit being continuous, we may apply A 5.2 to conclude that $\sum_j \xi_{nj} \xrightarrow{d}$ some ξ with $L_\xi = e^{-\psi}$, i.e. satisfying (6.2).

Since every bounded measure on R'_+ may be written in the form $\sum_j \mathfrak{g}P\xi_{nj}^{-1}$, it is easily seen that any α and λ with the stated properties may occur in (6.2). The infinite divisibility of the ξ in (6.2) thus follows by dividing this relation by an arbitrary $n \in N$. To see that conversely any infinitely divisible random variable ξ admits such a representation, note that if, for each $n \in N$, $\xi \overset{d}{=} \xi_{n1} + \cdots + \xi_{nn}$ with independent and identically distributed terms, then $L_{\xi_{n1}} = (L_\xi)^{1/n} \to 1$, which proves (6.4) and hence that the ξ_{nj} form a null-array. In the case of Z_+-valued random variables, the ξ_{nj} here are by definition Z_+-valued, and we get $\alpha = 0$ and $\lambda \in \mathfrak{M}(N)$ by (i) and (ii). Furthermore, (6.6) is in this case equivalent to the relation $\sum_j P\xi_{nj}^{-1} \xrightarrow{w} \lambda$ in $\mathfrak{M}(N)$. □

To get an extension to the multi-dimensional case, we need the following lemma which follows easily from the one-dimensional version just proved in combination with A 7.5 and Lemma 4.5.

Lemma 6.2. *Let* $\{\xi_{nj}\}$ *be a null-array of random measures on* $\mathfrak{S}$. *Then the sequence* $\left\{\sum_j \xi_{nj}\right\}$ *is relatively compact iff*

(i) $\limsup_{n\to\infty} \sum_j E[\xi_{nj}B;\, \xi_{nj}B < r] < \infty\,, \quad B \in \mathscr{B}\,, \quad r > 0\,,$

(ii) $\lim_{r\to\infty} \limsup_{n\to\infty} \sum_j P\{\xi_{nj}B > r\} = 0\,, \quad B \in \mathscr{B}\,.$

In the point process case, (i) *may be replaced by*

(i)′ $\limsup_{n\to\infty} \sum_j P\{\xi_{nj}B > 0\} < \infty\,, \quad B \in \mathscr{B}\,.$

Proof of Theorem 6.1 *in the k-dimensional case.* If $\mathfrak{S}$ consists of k points, then any random measure on $\mathfrak{S}$ may be identified with a random vector in R_+^k. Let us thus assume that $\{\xi_{nj}\}$ is a null-array of random vectors satisfying (i)—(iii) for some α and λ, where we choose for $\mathscr{J}$ the class of all one-point sets in $\mathfrak{S}$. It is then easily verified that

$$\sum_j \mathfrak{g}P(\xi_{nj}t)^{-1} \xrightarrow{w} \alpha t\delta_0 + \mathfrak{g}(\lambda\pi_t^{-1})\,, \qquad t \in R_+^k\,.$$

(Here the products are to be interpreted as inner products in R_+^k, and π_t denotes the projection $x \to xt$.) Applying the one-dimensional case, it follows that

$$E\exp\Big(-\sum_j \xi_{nj}t\Big) \to \exp\big(-\alpha t - \lambda(1 - e^{-\pi_t})\big)\,, \qquad t \in R_+^k\,,$$

and since the right-hand side here is a continuous function of t, we may apply A 5.2 to conclude that $\sum_j \xi_{nj} \xrightarrow{d}$ some ξ satisfying (6.2) in the form

$$-\log Ee^{-\xi t} = \alpha t + \lambda(1 - e^{-\pi_t})\,, \qquad t \in R_+^k\,.$$

In this representation, the uniqueness of α follows from the uniqueness in the one-dimensional case. As for λ, put $\mathbf{1} = (1, \dots, 1) \in R_+^k$ and note that

$$\int e^{-tx}(1 - e^{-x\mathbf{1}})\,\lambda(dx) = \log Ee^{-\xi t} - \log Ee^{-\xi(t+\mathbf{1})} - \alpha\mathbf{1}\,, \qquad t \in R_+^k\,.$$

Using A 5.1, it follows that the measure $(1 - e^{-\pi_1})\,\lambda$ and hence λ itself is unique in (6.2).

Suppose conversely that $\sum_j \xi_{nj} \xrightarrow{d}$ some ξ. Then it follows by the one-dimensional version that (ii) holds for some $\alpha \in R_+^k$ and that (iii) is fulfilled. Furthermore, it is seen from A 7.5 and Lemma 6.2 that $\sum_j P(\xi_{nj}t)^{-1}$ is vaguely relatively compact in $\mathfrak{M}(R_+^k\backslash\{0\})$, and so that any sequence $N' \subset N$ must contain a subsequence N'' such that (i) is satisfied for some λ as $n \to \infty$ through N''. Using the sufficiency and uniqueness assertions, it is seen that this λ is unique, so by A 1.2, (i) must hold for the original sequence. The proof may now be completed as before. □

To complete the proof of Theorem 6.1, we need the first part of the following lemma; the second part will not be needed until Section 11.

Lemma 6.3. *Let* ξ *be a random measure on* $\mathfrak{S}$ *and let* $\mathscr{J} \subset \mathscr{B}$ *be a DC-semiring. Then* ξ *is infinitely divisible iff* $(\xi I_1, \dots, \xi I_k)$ *is infinitely divisible for every* $k \in N$ *and* $I_1, \dots, I_k \in \mathscr{J}$. *In the point process case, it suffices that* $\xi I_1 + \cdots + \xi I_k$ *be infinitely divisible in* Z_+ *for every* $k \in N$ *and* $I_1, \dots, I_k \in \mathscr{J}$.

Proof. The necessity of the stated conditions is obvious. Conversely suppose that $(\xi I_1, \ldots, \xi I_k)$ is infinitely divisible for every $k \in N$ and $I_1, \ldots, I_k \in \mathcal{J}$. This condition extends immediately to the ring generated by $\mathcal{J}$. Since the class of infinitely divisible distributions in R_+^k is closed under weak convergence (as may easily be seen from A 5.2), we may apply A 2.2 and Lemma 1.2 to conclude that the assumption even extends to $\mathcal{B}$. Thus there exists for each $n \in N$ some family $P_{n, B_1, \ldots, B_k}$, $k \in N$, $B_1, \ldots, B_k \in \mathcal{B}$, of probability measures such that the corresponding L-transforms $\varphi_{B_1, \ldots, B_k}$ satisfy

$$(\varphi_{B_1, \ldots, B_k})^n = L_{\xi B_1, \ldots, \xi B_k}\,, \qquad k \in N\,, \qquad B_1, \ldots, B_k \in \mathcal{B}\,.$$

Using this relation, it is easy to verify that the measures $P_{B_1, \ldots, B_k}$ satisfy the conditions of Theorem 5.4, and hence that there exists some random measure ξ_n with these measures as finite-dimensional distributions. Furthermore, we may conclude from Theorem 3.1 that $(L_{\xi_n})^n = L_\xi$, and since n was arbitrary, ξ must indeed be infinitely divisible.

Let us next suppose that ξ is a point process such that $\xi I_1 + \cdots + \xi I_k$ is infinitely divisible in Z_+ for every $k \in N$ and $I_1, \ldots, I_k \in \mathcal{J}$. Since repetitions may occur among the sets I_j, it follows that, for fixed $k \in N$ and $I_1, \ldots, I_k \in \mathcal{J}$, $t_1 \xi I_1 + \cdots + t_k \xi I_k$ is infinitely divisible in Z_+ for any $t_1, \ldots, t_k \in Z_+$. Turning to rationals and using the closure property above, we may conclude that $t_1 \xi I_1 + \cdots + t_k \xi I_k$ is infinitely divisible in R_+ for every $\boldsymbol{t} = (t_1, \ldots, t_k) \in R_+^k$. For fixed $\boldsymbol{t}$ it thus follows by the one-dimensional version of Theorem 6.1 that

$$-\log E \exp\Bigl(-u \sum_j t_j \xi I_j\Bigr) = \alpha u + \int (1 - e^{-xu})\, \lambda(dx)\,, \qquad u \in R_+\,, \qquad (6.7)$$

for some $\alpha \in R_+$ and $\lambda \in \mathfrak{M}(R_+')$ with $\lambda \mathfrak{g} < \infty$. Letting $u \to \infty$ in (6.7), it is seen that $P\{\sum t_j \xi I_j = 0\} > 0$ iff $\alpha = 0$ and $\lambda R_+' < \infty$, and since these conditions hold for $\boldsymbol{t} = (1, \ldots, 1)$ by Theorem 6.1, we get for arbitrary $\boldsymbol{t}$, $P\{\sum t_j \xi I_j = 0\} \geqq P\{\sum \xi I_j = 0\} > 0$, and therefore $\alpha = 0$ and $\lambda R_+' < \infty$ in general. But then $\sum t_j \xi I_j$ is a compound Poisson variable with compounding distribution $\lambda / \lambda R_+'$ (provided we exclude the trivial case $\lambda = 0$), so λ must be supported by the set $\{\boldsymbol{t}z : z \in Z_+^k \setminus \{0\}\}$, and if the t_j are assumed to be rationally independent, we may write (6.7) in the form

$$-\log E \exp\Bigl(-u \sum_j t_j \xi I_j\Bigr) = \sum_z (1 - e^{-u\boldsymbol{t}z})\, \lambda\{\boldsymbol{t}z\}\,, \qquad u \in R_+\,.$$

(This may also be seen more directly from Theorem 6.8 below.)

According to the k-dimensional version of Theorem 6.1, there exists some infinitely divisible Z_+^k-valued random vector η satisfying

$$-\log E e^{-v\eta} = \sum_z (1 - e^{-vz})\, \lambda\{\boldsymbol{t}z\}\,, \qquad v \in R_+^k\,,$$

and putting $v = u\boldsymbol{t}$, we may conclude from A 5.1 that $\boldsymbol{t}\eta \stackrel{d}{=} \sum_j t_j \xi I_j$. By the rational independence of the t_j, this implies $\eta \stackrel{d}{=} (\xi I_1, \ldots, \xi I_k)$, which proves that the latter vector is infinitely divisible. Thus it follows from the first part of the lemma that ξ is infinitely divisible as a random measure. Since moreover all the ξI are infinitely divisible in Z_+, we may argue as in the proof of Lemma 1.5 to show that ξ is even infinitely divisible as a point process. □

Proof of Theorem 6.1 in the general case. Suppose that ξ is an a.s. bounded infinitely divisible random measure. Then $(\xi B_1, \ldots, \xi B_k)$ is infinitely divisible for every $k \in N$ and $B_1, \ldots, B_k \in \mathscr{S}$, so according to the k-dimensional version of the theorem, there exist unique vectors $\alpha_{B_1, \ldots, B_k} \in R_+^k$ and measures $\lambda_{B_1, \ldots, B_k} \in \mathfrak{M}(R_+^k \backslash \{0\})$ satisfying

$$-\log E \exp\left(-\sum_{j=1}^{k} t_j \xi B_j\right) = \alpha_{B_1, \ldots, B_k} \boldsymbol{t} + \lambda_{B_1, \ldots, B_k}(1 - e^{-\pi_t}) \tag{6.8}$$

for any $\boldsymbol{t} = (t_1, \ldots, t_k) \in R_+^k$. Choosing all but one of the t_j equal to 0 yields

$$\alpha_{B_1, \ldots, B_k} = (\alpha_{B_1}, \ldots, \alpha_{B_k}), \qquad k \in N, \qquad B_1, \ldots, B_k \in \mathscr{S}, \tag{6.9}$$

while choosing exactly one t_j equal to 0 shows that the measures $\lambda_{B_1, \ldots, B_k}$ are consistent (in the obvious sense). Let us now consider disjoint $B, C \in \mathscr{S}$ and define $\lambda'_{B, C, B \cup C} \in \mathfrak{M}(R_+^3 \backslash \{0\})$ by

$$\lambda'_{B, C, B \cup C} A \equiv \lambda_{B, C}\{(x, y) \colon (x, y, x + y) \in A\},$$

where A denotes an arbitrary BOREL set. By (6.8) and (6.9), we then obtain for any $s, t, u \in R_+$

$$\begin{aligned} s\alpha_B + t\alpha_C + u\alpha_{B \cup C} &+ \int (1 - e^{-sx - ty - uz}) \lambda_{B, C, B \cup C}(dx\, dy\, dz) \\ &= -\log E e^{-s\xi B - t\xi C - u\xi(B \cup C)} = -\log E e^{-(s+u)\xi B - (t+u)\xi C} \\ &= (s + u)\alpha_B + (t + u)\alpha_C + \int (1 - e^{-(s+u)x - (t+u)y})\lambda_{B, C}(dx\, dy) \\ &= s\alpha_B + t\alpha_C + u(\alpha_B + \alpha_C) + \int (1 - e^{-sx - ty - uz}) \lambda'_{B, C, B \cup C}(dx\, dy\, dz), \end{aligned}$$

and it follows from the uniqueness in the 3-dimensional case that

$$\alpha_{B \cup C} = \alpha_B + \alpha_C, \qquad \lambda_{B, C, B \cup C} = \lambda'_{B, C, B \cup C}. \tag{6.10}$$

Here the latter relation yields

$$\begin{aligned} \lambda_{B, C, B \cup C}\{(x, y, z) \colon x + y \neq z\} &= \lambda'_{B, C, B \cup C}\{(x, y, z) \colon x + y \neq z\} \\ &= \lambda_{B, C}\{(x, y) \colon x + y \neq x + y\} = 0. \end{aligned} \tag{6.11}$$

Next suppose that $B_1, B_2, \ldots \in \mathscr{S}$ with $B_n \downarrow \emptyset$, and conclude from (6.8) with $k = 1$ and $t_1 = 1$ that

$$\alpha_{B_n} \downarrow 0; \qquad \lambda_{B_n}(\varepsilon, \infty) \downarrow 0, \qquad \varepsilon > 0. \tag{6.12}$$

Combining (6.10) and (6.12), it is seen that the set function α_B, $B \in \mathscr{S}$, is an element of $\mathfrak{M}$. As for the family of λ-measures, note that (6.11) and (6.12) remain true for the measures $\lambda''_{B_1, \ldots, B_k}$ defined by

$$\lambda''_{B_1, \ldots, B_k} A = \int (1 - e^{-x}) \lambda_{\mathfrak{S}, B_1, \ldots, B_k}(dx \times A), \quad k \in N, \quad B_1, \ldots, B_k \in \mathscr{S}. \tag{6.13}$$

Since $\lambda''_{B_1, \ldots, B_k} R_+^k = \lambda_{\mathfrak{S}}\mathfrak{g} < \infty$, there exists by Theorem 5.4 some unique bounded measure λ'' on $\mathfrak{M} \backslash \{0\}$ satisfying

$$\lambda'' \pi^{-1}_{B_1, \ldots, B_k} = \lambda''_{B_1, \ldots, B_k}, \qquad k \in N, \qquad B_1, \ldots, B_k \in \mathscr{S}. \tag{6.14}$$

Defining the measure λ on $\mathfrak{M} \backslash \{0\}$ by

$$\lambda(d\mu) \equiv (1 - e^{-\mu \mathfrak{S}})^{-1} \lambda''(d\mu),$$

it follows by (6.13) and (6.14) that

$$\begin{aligned}\lambda\pi_{B_1,\ldots,B_k}^{-1}A &= \int_{(\mu B_1,\ldots,\mu B_k)\in A} (1-e^{-\mu\mathfrak{S}})^{-1}\,\lambda''(d\mu)\\ &= \int (1-e^{-x})^{-1}\,\lambda''_{\mathfrak{S},B_1,\ldots,B_k}(dx\times A)\\ &= \int (1-e^{-x})^{-1}\,(1-e^{-y})\,\lambda_{\mathfrak{S},\mathfrak{S},B_1,\ldots,B_k}(dx\,dy\times A)\\ &= \lambda_{\mathfrak{S},\mathfrak{S},B_1,\ldots,B_k}\,(R_+^2\times A) = \lambda_{\mathfrak{S},B_1,\ldots,B_k}(R_+\times A)\,,\end{aligned}$$

and inserting this into (6.8) yields (6.2) for simple functions, and hence in general by monotone convergence.

For general (possibly unbounded) ξ, it is easily seen from A 6.1 and the Borel-Cantelli lemma that there exists some continuous and strictly positive function $h\in\mathcal{F}$ with $\xi h<\infty$ a.s. Since $h\xi$ is then an a.s. bounded infinitely divisible random measure, there exist as above some unique measures α_h and λ_h satisfying

$$\begin{aligned}-\log Ee^{-\xi f} &= -\log Ee^{-(h\xi)(f/h)} = \alpha_h(f/h) + \int (1-e^{-\mu(f/h)})\,\lambda_h(d\mu)\\ &= (h^{-1}\alpha_h)\,f + \int (1-e^{-(h^{-1}\mu)f})\,\lambda_h(d\mu)\end{aligned}$$

for arbitrary $f\in\mathcal{F}$. Hence (6.2) holds in this case with $\alpha=h^{-1}\alpha_h$ and $\lambda(d\mu)\equiv\lambda_h\{m: h^{-1}m\in d\mu\}$. It is easily verified that these measures α and λ have the stated properties. Conversely, it follows from Theorem 5.4 that any α and λ with these properties define the distribution of some random measure ξ satisfying (6.2). In particular, these distributions must be infinitely divisible.

If $\{\xi_{nj}\}$ is a null-array of random measures such that $\sum_j \xi_{nj}\xrightarrow{d}$ some ξ, then it is easily seen from Lemma 6.3 that ξ is infinitely divisible. Thus the convergence criteria follow from those in the finite-dimensional case. The results for point processes are easily decuded from (i)—(iii). □

We shall call (6.2) the *canonical representation* of L_ξ, while α and λ will be called the *canonical measures* of ξ. It will often be convenient to write $I(\alpha,\lambda)$ and $I(\lambda)$ for the corresponding classes of infinitely divisible random measures and point processes respectively.

Corollary 6.4. *Let $\{\xi_{nj}\}$ be a null-array of random measures on $\mathfrak{S}$ and suppose that $\eta_n\in I(0,\sum_j P\xi_{nj}^{-1})$, $n\in N$. Then $\sum_j \xi_{nj}\xrightarrow{d}$ some ξ iff $\eta_n\xrightarrow{d}$ some η, and in that case $\xi\overset{d}{=}\eta$.*

Proof. Suppose that $\eta_n\xrightarrow{d}\eta$. Then $\eta\in$ some $I(\alpha,\lambda)$ by Lemma 6.3, and applying Theorem 4.2 to the representation (6.2) for η_n and η, it is seen that (6.3) holds. By Theorem 6.1, this yields $\xi_n\xrightarrow{d}\eta$. A similar argument proves the converse assertion. □

Cluster representations

The canonical representation (6.2) admits the following useful and illuminating interpretations.

Lemma 6.5. *Let $\xi\in I(\alpha,\lambda)$ on $\mathfrak{S}$ and let η be a Poisson process on $\mathfrak{M}\backslash\{0\}$ with intensity λ. Then*

$$\xi\overset{d}{=}\alpha+\int\mu\eta(d\mu)\,. \tag{6.15}$$

Proof. Let ζ denote the right-hand side of (6.15). For fixed $B \in \mathscr{B}$, the measurability of ζB follows from Lemma 1.3 (in the version for measures on $\mathfrak{M}\backslash\{0\}$) and the fact that the mapping $\pi_B \colon \mu \to \mu B$ is measurable. Furthermore, it is seen from Table 1 and (6.2) that

$$Ee^{-\zeta f} = Ee^{-\alpha f - \int \mu f \eta(d\mu)} = e^{-\alpha f} Ee^{-\eta \pi f} = e^{-\alpha f} \exp\left(-\lambda(1 - e^{-\pi f})\right) = Ee^{-\xi f}, \quad f \in \mathscr{F}. \tag{6.16}$$

It follows in particular that $\zeta B < \infty$ a.s. for all $B \in \mathscr{B}$, and so, by Lemma 1.6, that ζ is a random measure on $\mathfrak{S}$. But then Theorem 3.1 applies, and we may conclude from (6.16) that $\zeta \stackrel{d}{=} \xi$. □

Lemma 6.6. *Let $\xi \in I(\alpha, \lambda)$ on $\mathfrak{S}$, and let $M_1, M_2, \ldots \in \mathscr{M}$ form a disjoint partition of $\mathfrak{M}\backslash\{0\}$ with $\lambda M_n < \infty$, $n \in N$. Further suppose that, for each $n \in N$, ν_n is a Poissonian random variable with mean λM_n while $\xi_{n1}, \xi_{n2}, \ldots$ are random measures on $\mathfrak{S}$ with the common distribution $M_n\lambda/\lambda M_n$. Letting all these random elements be independent, we have*

$$\xi \stackrel{d}{=} \alpha + \sum_{n=1}^{\infty} \sum_{j=1}^{\nu_n} \xi_{nj}. \tag{6.17}$$

Proof. Writing ζ for the right-hand side of (6.17), we get for any $f \in \mathscr{F}$

$$\begin{aligned} L_\xi(f) &= e^{-\alpha f} \prod_{n=1}^{\infty} \left\{ e^{-\lambda M_n} \sum_{k=0}^{\infty} \frac{(\lambda M_n)^k}{k!} \left(\frac{(M_n\lambda)\, e^{-\pi f}}{\lambda M_n} \right)^k \right\} \\ &= e^{-\alpha f} \prod_{n=1}^{\infty} \{ e^{-\lambda M_n} \exp\left((M_n\lambda)\, e^{-\pi f} \right) \} = e^{-\alpha f} \prod_{n=1}^{\infty} \exp\{ -(M_n\lambda)\,(1 - e^{-\pi f}) \} \\ &= e^{-\alpha f} \exp\left\{ -\sum_{n=1}^{\infty} (M_n\lambda)\,(1 - e^{-\pi f}) \right\} = \exp\{-\alpha f - \lambda\,(1 - e^{-\pi f})\} = L_\xi(f) \end{aligned}$$

and it follows by Lemma 1.6 and Theorem 3.1 that $\zeta \stackrel{d}{=} \xi$. □

The canonical measures

We shall now use Lemma 6.6 to examine the relationship between the infinitely divisible random measures and their canonical measures.

Theorem 6.7. *Let $\xi \in I(\alpha, \lambda)$ and let $a > 0$ be fixed. Then $\xi_a^* = 0$ a.s. iff*

(i) $\alpha_a^* = 0$,
(ii) $\lambda\{\mu_a^* \neq 0\} = 0$,
(iii) $\lambda\{\mu\{s\} > 0\} = 0, \quad s \in \mathfrak{S}$.

In particular, ξ is a.s. diffuse iff $\alpha \in \mathfrak{M}_d$ and $\lambda(\mathfrak{M}_d)^c = 0$.

Proof. We may clearly identify ξ with the right-hand side of (6.17). Let us first suppose that $\xi_a^* = 0$ a.s. Then (i) is obvious while (ii) follows from the fact that, for each $n \in N$, $P\{\nu_n \geqq 1\} > 0$ and hence by (6.17) $(\xi_{n1})_a^* = 0$ a.s. Finally, we obtain for any $s \in \mathfrak{S}$ and $k, n \in N$

$$\begin{aligned} 0 = P\{\xi_a^* \neq 0\} \geqq P\{\xi\{s\} \geqq a\} &\geqq P\{\nu_n \geqq k; \xi_{nj}\{s\} \geqq a/k,\ j = 1, \ldots, k\} \\ &= P\{\nu_n \geqq k\} \left(P\{\xi_{n1}\{s\} \geqq a/k\}\right)^k, \end{aligned}$$

which yields $\xi_{n1}\{s\} < a/k$ a.s. since the first factor on the right is positive. Since k was arbitrary, this shows that $\xi_{n1}\{s\} = 0$ a.s., and (iii) follows.

Conversely, suppose that (i)—(iii) hold. Since by (ii) $(\xi_{nj})_a^* = 0$ a.s. for all n and j, it suffices to prove that, with probability 1, no two measures among α and the ξ_{nj} can have atom positions in common. Since each measure can have at most countably many atoms, this follows by FUBINI's theorem from the fact that, according to (iii), $\xi_{nj}B = 0$ a.s. for every fixed countable set $B \subset \mathfrak{S}$. □

To state the next result, we need some further notation. For any measure λ on $\mathfrak{M}$, let $A(\lambda)$ be the set of atom positions of λ and denote by $S(\lambda)$ the vague support of λ. This means that $\mu \in A(\lambda)$ iff $\lambda\{\mu\} > 0$, while $\mu \in S(\lambda)$ iff $\lambda G > 0$ for all vaguely open sets G containing μ. Given any set $M \subset \mathfrak{M}$, we further write $\mathcal{H}M$ for the additive semigroup in $\mathfrak{M}$ which is generated by $M \cup \{0\}$.

Theorem 6.8. *Let $\xi \in I(\alpha, \lambda)$ on $\mathfrak{S}$. Then*

$$A(P\xi^{-1}) = \begin{cases} \alpha + \mathcal{H}A(\lambda)\,, & \lambda(\mathfrak{M}\setminus\{0\}) < \infty\,, \\ \emptyset, & \lambda(\mathfrak{M}\setminus\{0\}) = \infty\,, \end{cases} \tag{6.18}$$

$$S(P\xi^{-1}) = \alpha + \big(\mathcal{H}S(\lambda)\big)^{-}\,. \tag{6.19}$$

Proof. For bounded λ, (6.18) follows directly from Lemma 6.6 by choosing $M_1 = \mathfrak{M}\setminus\{0\}$, $M_2 = M_3 = \cdots = \emptyset$. Next suppose that $\lambda(\mathfrak{M}\setminus\{0\}) = \infty$. Choose a strictly positive function $f \in \mathcal{F}$ such that $\xi f < \infty$ a.s. (use the BOREL-CANTELLI lemma), and note that $\lambda\pi_f^{-1}R_+' = \lambda\{\mu f > 0\} = \lambda(\mathfrak{M}\setminus\{0\}) = \infty$. If (6.18) is known to be true for random variables, we get $P\{\xi = \mu\} = P\{\xi f = \mu f\} = 0$, $\mu \in \mathfrak{M}$, so $P\xi^{-1}$ must be diffuse. We may thus assume from now on that $\mathfrak{S} = R_+$. Since λ is unbounded, there exists for each $n \in N$ some decomposition $\lambda = \lambda_n + \lambda_n'$ such that $\lambda_n'R_+ = \lambda_n'[0, 1/n] = \log 2$, and we may write $\xi \overset{d}{=} \xi_n + \eta_n$ where $\xi_n \in I(\alpha, \lambda_n)$ and $\eta_n \in I(0, \lambda_n')$ are independent. Letting $x \geqq 0$ and $\varepsilon > 0$, we get

$$\begin{aligned} P\{\xi = x\} &\leqq P\{\xi_n = x\}\ P\{\eta_n = 0\} + P\{x - \varepsilon \leqq \xi_n < x\}\ P\{\eta_n > 0\} \\ &\quad + P\{\eta_n > \varepsilon\} = \frac{1}{2} P\{x - \varepsilon \leqq \xi_n \leqq x\} + P\{\eta_n > \varepsilon\}\,, \end{aligned}$$

and since $\xi_n \xrightarrow{d} \xi$ and $\eta_n \xrightarrow{d} 0$, we may conclude from A 4.1 that $P\{\xi = x\} \leqq \frac{1}{2} P\{x - \varepsilon \leqq \xi \leqq x\}$. As $\varepsilon \to 0$ this yields $P\{\xi = x\} = 0$, and since x was arbitrary, we obtain $A(P\xi^{-1}) = \emptyset$, which completes the proof of (6.18).

To prove (6.19), note first that $\xi \geqq \alpha$ a.s. by Lemma 6.6. Thus $P(\xi - \alpha)^{-1}$ is well-defined, and the conditions $\mu \in S(P\xi^{-1}) - \alpha$ and $\mu \in S\big(P(\xi - \alpha)^{-1}\big)$ are seen to be equivalent since both mean that

$$P \bigcap_{j=1}^{k} \{|(\xi - \alpha - \mu)\, f_j| < \varepsilon\} > 0\,, \qquad \varepsilon > 0\,, \quad k \in N\,, \quad f_1, \ldots, f_k \in \mathcal{F}_c\,.$$

It is therefore enough to prove (6.19) in the case $\alpha = 0$. Throughout the proof, we shall further assume that ϱ is a fixed complete metric in $\mathfrak{M}$ (cf. A 7.7) and that $\lambda \neq 0$. (If $\lambda = 0$, (6.19) reduces to a triviality.)

Let us first assume that λ is bounded. In this case we may write $\xi \overset{d}{=} \xi_1 + \cdots + \xi_\nu$, where ν is POISSONian with mean $\lambda(\mathfrak{M}\setminus\{0\})$ while the ξ_j are independent of ν

processes by KERSTAN, MATTHES and MECKE [44], p. 95, and in Theorem 6.8, the first part is due (for random variables) to DOEBLIN [13], while the second part is new.

A different approach to infinite divisibility (for point processes) is suggested in [44], where (6.17) playes the role of a basis for the whole theory. See also [44], pp. 287 ff, for a complete theory of cluster representations. We finally point out that infinite divisibility w.r.t convolution (rather than addition) of a.s. bounded random measures is studied in [39].

Exercises

6.1. Give a direct proof of Theorem 6.1 in the particular case of Z_+-valued random variables, using generating functions.

6.2. Let $\{\xi_{nj}\}$ be a null-array of random measures on $\mathfrak{S}$. Show that $\sum_i \xi_{nj} \xrightarrow{d}$ some ξ iff $\sum_j \mathfrak{g}P(\xi_{nj}f)^{-1}$ converges weakly for every $f \in \mathcal{F}_c$. (Hint: Use Lemma 5.1.)

6.3. Let $\xi_1, \xi_2, \ldots$ be independent random measures on $\mathfrak{S}$. Show that $\sum_j \xi_j$ represents a random measure iff $\sum_j (1 - Ee^{-\xi_j B}) < \infty$, $B \in \mathcal{B}$. (Hint: Use Lemma 1.6 and Theorem 6.1. Cf. [15], p. 317.)

6.4. Show that the class of infinitely divisible random measures is closed. Let $\xi \in I(\alpha, \lambda)$ and $\xi_n \in I(\alpha_n, \lambda_n)$, $n \in N$, on $\mathfrak{S}$, and let $\mathcal{J} \subset \mathcal{B}_\alpha \cap \mathcal{B}_\lambda$ be a DC-semiring. Prove that $\xi_n \xrightarrow{d} \xi$ iff

(i) $\lambda_n \pi^{-1}_{I_1,\ldots,I_k} \xrightarrow{v} \lambda \pi^{-1}_{I_1,\ldots,I_k}$ in $\mathfrak{M}(R^k_+/\{0\})$, $k \in N$, $I_1, \ldots, I_k \in \mathcal{J}$,

(ii) $\lim_{\varepsilon \to 0} \operatorname{limsi}_{n \to \infty} (\alpha_n I + \int_{x < \varepsilon} x \lambda_n \{\mu I \in dx\}) = \alpha I$, $I \in \mathcal{J}$,

(iii) $\lim_{r \to \infty} \limsup_{n \to \infty} \lambda_n \{\mu I > r\} = 0$, $I \in \mathcal{J}$,

and in the point process case iff

$$\lambda_n \pi^{-1}_{I_1,\ldots,I_k} \xrightarrow{w} \lambda \pi^{-1}_{I_1,\ldots,I_k} \text{ in } \mathfrak{M}(Z^k_+/\{0\}),\ k \in N,\ I_1, \ldots, I_k \in \mathcal{J}\,.$$

(Cf. JIŘINA [29] and NAWROTZKI [63].)

6.5. Let $\xi_n \in I(\alpha_n, \lambda_n)$ on $\mathfrak{S}$, $n \in N$. Show that $\{\xi_n\}$ is relatively compact iff, for every fixed $B \in \mathcal{B}$, the sequences $\{\alpha_n B\}$ and $\{\lambda_n \pi_B^{-1} \mathfrak{g}\}$ are bounded while $\lim_{r \to \infty} \limsup_{n \to \infty} \lambda_n \{\mu B > r\} = 0$.

6.6. Let $\xi \in I(\alpha, \lambda)$ on $\mathfrak{S}$. Show that $E\xi = \alpha + \int \mu \lambda(d\mu)$. (Cf. [44], p. 93. Hint: Use Lemma 6.5.)

6.7. Let $\xi \in I(\alpha, \lambda)$ on $\mathfrak{S}$ and let $f \in \mathcal{F}$. Show that $\xi f < \infty$ a.s. iff $\alpha f < \infty$, $\lambda\{\mu f = \infty\} = 0$ and $\lambda(1 - e^{-\pi_f}) < \infty$. (Cf. [44], p. 94. Hint: Let $t \to 0$ in $L_{\xi f}(t)$.)

6.8. Let $\xi \in I(\alpha, \lambda)$ on $\mathfrak{S}$ and let $f \in \mathcal{F}$ be such that $(f\xi)\, B < \infty$ a.s., $B \in \mathcal{B}$. Show that $f\xi \in I(f\alpha, \lambda_f)$, where λ_f is the restriction to $\mathfrak{M} \backslash \{0\}$ of the image of λ under the mapping $\mu \to f\mu$.

6.9. Let $\xi \in I(\alpha, \lambda)$ on $\mathfrak{S}$. Show that ξ is a point process iff $\alpha \in \mathfrak{N}$ and $\lambda \mathfrak{N}^c = 0$. (Hint: Use A 7.4 and Theorem 6.8.)

6.10. Show that a point process ξ on $\mathfrak{S}$ is infinitely divisible as a point process iff it is infinitely divisible as a random measure and $P\{\xi B = 0\} > 0$, $B \in \mathcal{B}$. (Hint: Cf. Exercise 6.9 and Theorem 6.8.)

6.11. Let ξ be a point process on $\mathfrak{S}$ and let $\mathcal{J} \subset \mathcal{B}$ be a DC-semiring. Show that ξ is infinitely divisible as a random measure iff $\xi I_1 + \cdots + \xi I_k$ is infinitely divisible in R_+ for every $k \in N$ and $I_1, \ldots, I_k \in \mathcal{J}$. (Hint: Proceed as in the proof of Lemma 6.3, using Theorem 6.8.)

6.12. Let $\xi \in I(\alpha, \lambda)$ on $\mathfrak{S}$. Show that $P\xi^{-1}$ is purely atomic iff λ is bounded and purely atomic. (Hint: Proceed as in the proof of (6.18).)

6.13. Show that a point process ξ on $\mathfrak{S}$ is infinitely divisible iff ξf is infinitely divisible in R_+ and $P\{\xi f = 0\} > 0$ for every $f \in \mathcal{F}_c$. (Hint: Use the second part of Lemma 6.3 together with Theorem 6.8.)

6.14. Let $\mathfrak{S}$ be countable and compact. Show that there exists some fixed function $f \in \mathcal{F}$ such that a point process ξ on $\mathfrak{S}$ is infinitely divisible as a random measure iff ξf is infinitely divisible in R_+. (Hint: Cf. Exercises 3.7 and 6.12.)

6.15. Show that if $\mathfrak{S}$ contains more than one point, then there exists a point process ξ on $\mathfrak{S}$ such that ξ is not infinitely divisible and still ξB is infinitely divisible as a random variable in Z_+ for every $B \in \mathcal{B}$. (Hint: Use the idea of Exercise 3.8, and apply Theorem 6.8.)

7. Infinite divisibility, special cases

The results of Section 6 will now be improved in important special cases.

Independent increments

Say that the random measure ξ on $\mathfrak{S}$ has a *fixed atom at* $s \in \mathfrak{S}$ if $\xi\{s\} \overset{d}{\neq} 0$. We prove first that a random measure with independent increments is infinitely divisible, apart from at most countably many fixed atoms.

Theorem 7.1. *A random measure ξ on $\mathfrak{S}$ has independent increments iff it can be written in the form*

$$\xi = \eta + \sum_{j=1}^{k} \beta_j \delta_{t_j} \tag{7.1}$$

for some fixed $k \in Z_+ \cup \{\infty\}$ and $t_1, t_2, \ldots \in \mathfrak{S}$, some infinitely divisible random measure η with independent increments and without fixed atoms, and some R_+-valued random variables $\beta_j \overset{d}{\neq} 0$, $j = 1, 2, \ldots$, which are mutually independent and independent of η. In this case, the decomposition (7.1) is a.s. unique apart from the order of terms.

Proof. Like any random measure, ξ can have at most countably many fixed atoms. To see this, it clearly suffices to show that, for any fixed $B \in \mathcal{B}$ and $\varepsilon > 0$, the relation $P\{\xi\{s\} \geqq \varepsilon\} \geqq \varepsilon$ can hold for at most finitely many $s \in B$. Assuming on the contrary that this relation is fulfilled for infinitely many distinct points $s_1, s_2, \ldots \in B$, we reach by FATOU's lemma the contradiction

$$0 = P\{\xi B = \infty\} \geqq P \limsup_{n\to\infty} \{\xi\{s_n\} \geqq \varepsilon\} \geqq \limsup_{n\to\infty} P\{\xi\{s_n\} \geqq \varepsilon\} \geqq \varepsilon .$$

Subtracting the fixed atoms $\beta_1\delta_{t_1}, \beta_2\delta_{t_2}, \ldots$ from ξ, we are left with a random measure η with independent increments which has no fixed atoms, and it remains to prove that η is infinitely divisible. For this purpose, let $B \in \mathcal{B}$ be arbitrary and let $\{B_{nj}\} \subset \mathcal{B}$ be a null-array of partitions of B. Then $\{\eta B_{nj}\}$ is a null-array of random variables. In fact, suppose on the contrary that $P\{\eta B_{nj_n} \geqq \varepsilon\} \geqq \varepsilon$, $n \in N'$, for some $\varepsilon > 0$, some sequence $N' \subset N$ and some indices j_n, $n \in N'$. Proceeding as in the proof of Lemma 2.2, we may then conclude that there exists some point $s \in B^-$ such that $P\{\eta G \geqq \varepsilon\} \geqq \varepsilon$ for every open set $G \in \mathcal{B}$ containing s. Since $G_n \downarrow \{s\}$ implies $\eta G_n \downarrow \eta\{s\}$ a.s., it follows by A 4.1 that $P\{\eta\{s\} \geqq \varepsilon\} \geqq \varepsilon$, which contradicts the fact that η has no fixed atoms.

Since $\eta B = \sum_j \eta B_{nj}$, $n \in N$, we may now conclude from Theorem 6.1 that ηB is infinitely divisible. A similar argument shows more generally that $(\eta B_1, \dots, \eta B_k)$ is infinitely divisible for arbitrary $k \in N$ and $B_1, \dots, B_k \in \mathscr{B}$, and by Lemma 6.3, this yields the infinite divisibility of η itself. □

We are now going to consider the special case of infinite divisibility in detail. In the present case, surprisingly weak conditions are needed for convergence, as compared with those in Theorem 6.1. Given any class $\mathscr{I} \subset \mathscr{B}$, we shall mean by $\mathscr{I} \gg \mathscr{B}$ that every set $B \in \mathscr{B}$ is contained in some $I \in \mathscr{I}$.

Theorem 7.2. *The relation*

$$-\log Ee^{-\xi f} = \alpha f + \int (1 - e^{-xf(t)})\,\gamma(dx\,dt)\,, \qquad f \in \mathscr{F}\,, \tag{7.2}$$

defines a unique correspondence between the distributions of all infinitely divisible random measures ξ on $\mathfrak{S}$ with independent increments and the class of all pairs (α, γ), where $\alpha \in \mathfrak{M}$ while $\gamma \in \mathfrak{M}(R_+' \times \mathfrak{S})$ with $\gamma(\cdot \times B)\,\mathfrak{g} < \infty$, $B \in \mathscr{B}$. If $\{\xi_{nj}\}$ is a null-array of random measures on $\mathfrak{S}$ and if $\mathscr{I} \subset \mathscr{B}_\xi$ is a DC-semiring with $\mathscr{I} \gg \mathscr{B}$, then $\sum_j \xi_{nj} \xrightarrow{d} \xi$ iff $\sum_j \xi_{nj} I \xrightarrow{d} \xi I$, $I \in \mathscr{I}$, i.e. iff

$$\sum_j \mathfrak{g} P(\xi_{nj} I)^{-1} \xrightarrow{w} \alpha I \delta_0 + \mathfrak{g}\gamma(\cdot \times I)\,, \qquad I \in \mathscr{I}\,. \tag{7.3}$$

In the point process case we have $\alpha = 0$ while γ is confined to $N \times \mathfrak{S}$, and in this case (7.3) *is equivalent to*

$$\sum_j P(\xi_{nj} I)^{-1} \xrightarrow{w} \gamma(\cdot \times I) \text{ in } \mathfrak{M}(N)\,, \qquad I \in \mathscr{I}\,.$$

Our proof is based on the following lemma. Let us say that a measure is *degenerate* if all its mass is confined to one single point.

Lemma 7.3. *Let $\xi \in I(\alpha, \lambda)$ on $\mathfrak{S}$. Then ξ has independent increments iff λ is confined to the set if degenerate measures in $\mathfrak{M}\backslash\{0\}$.*

Proof. Suppose that λ has the stated property, and consider arbitrary $k \in N$ and disjoint $B_1, \dots, B_k \in \mathscr{B}$. By Theorem 6.1 we get for any $t_1, \dots, t_k \in R_+$

$$\begin{aligned}
-\log E \exp\Big(-\sum_j t_j \xi B_j\Big) &= \sum_j t_j \alpha B_j + \int \Big(1 - \exp\Big(-\sum_j t_j \mu B_j\Big)\Big)\lambda(d\mu) \\
&= \sum_j t_j \alpha B_j + \sum_j \int_{\mu B_j > 0} (1 - \exp(-t_j \mu B_j))\,\lambda(d\mu) \\
&= \sum_j \{t_j \alpha B_j + \int (1 - \exp(-t_j \mu B_j))\,\lambda(d\mu)\} = -\log \prod_j E \exp(-t_j \xi B_j),
\end{aligned}$$

and it follows by A 5.1 that $\xi B_1, \dots, \xi B_k$ are independent.

Conversely, suppose that ξ has independent increments, and consider any two disjoint sets $B, C \in \mathscr{B}$. Then the above calculations yield

$$\int \{(1 - e^{-\mu B}) + (1 - e^{-\mu C}) - (1 - e^{-\mu B - \mu C})\}\,\lambda(d\mu) = 0\,. \tag{7.4}$$

Now the function $1 - e^{-x}$ is strictly concave on R_+, so we have

$$1 - e^{-x-y} \leqq (1 - e^{-x}) + (1 - e^{-y})\,, \qquad x, y \in R_+\,,$$

with equality iff $x \wedge y = 0$. Thus the integrand in (7.4) must be zero a.e. λ, and we get $\mu B \wedge \mu C = 0$ a.e. λ. This result implies more generally that, for arbitrary $k \in N$ and disjoint $B_1, \dots, B_k \in \mathscr{B}$, at most one of the quantities $\mu B_1, \dots, \mu B_k$ can be positive a.e. λ. We may now apply Lemma 2.2 to conclude that $B\mu$ is degenerate a.e. λ for any fixed $B \in \mathscr{B}$. By A 6.1, this result remains true for $B = \mathfrak{S}$. □

Proof of Theorem 7.2. Suppose that ξ is infinitely divisible with independent increments. Let M be the set of degenerate measures in $\mathfrak{M}\backslash\{0\}$, and for $\mu \in M$, let t_μ denote the unique atom position of μ. From Lemma 2.3 it is seen that $M \in \mathscr{M}$ and that $\mu \to (\mu S, t_\mu)$ is a measurable mapping of M into $R'_+ \times \mathfrak{S}$. Let γ be the image of λ under this mapping. Then (7.2) is obtained by a corresponding transformation of the integral in (6.2), and since clearly

$$\gamma(dx \times B) = \lambda\{\mu B \in dx\}\,, \qquad x > 0\,, \qquad B \in \mathscr{B}\,, \tag{7.5}$$

we have $\gamma(\cdot \times B)\,\mathfrak{g} = (\lambda \pi_B^{-1})\,\mathfrak{g} < \infty$, $B \in \mathscr{B}$. To see that the representation (7.2) is unique, note that (7.2) may be transformed into (6.2) with λ now denoting the image of γ under the inverse mapping $(x, t) \to x\delta_t$. Since (7.5) remains true, and hence γ is uniquely determined by λ, the asserted uniqueness follows from the corresponding statement in Theorem 6.1. The simplifications stated for the point process case follow easily by (7.5) and Theorem 6.1.

As for the convergence assertion, the necessity of (7.3) follows from (7.5) and Theorem 6.1. Conversely, suppose that (7.3) holds. In proving conditions (i) — (iii) of Theorem 6.1, the only non-trivial part is to verify (i) for $k > 1$, and for this purpose it is clearly enough to show that, for any disjoint sets $J_1, J_2 \in \mathscr{J}$ and any $\varepsilon > 0$,

$$\lim_{n\to\infty} \sum_j P\{\xi_{nj}J_1 \wedge \xi_{nj}J_2 > \varepsilon\} = 0\,, \qquad \varepsilon > 0\,. \tag{7.6}$$

By the assumptions on $\mathscr{J}$, we may choose some $I \in \mathscr{J}$ admitting a disjoint partition into sets $I_1, \ldots, I_m \in \mathscr{J}$ with the property that J_1 and J_2 are both unions of sets in $\{I_1, \ldots, I_m\}$. In order to prove (7.6), it suffices by condition (iii) of Theorem 6.1 to show that

$$\lim_{n\to\infty} \sum_j P\{(\xi_{nj}I_1, \ldots, \xi_{nj}I_m) \in C\} = 0 \tag{7.7}$$

for every compact set $C \in R_+^m$ which is bounded away from all the coordinate axes. Now every point in C is clearly contained in the set

$$\{(x_1, \ldots, x_m) \in R_+^m : \max x_k < t,\ \sum x_k > t\}$$

for some $t > 0$ with $\gamma(\{t\} \times I) = 0$, and since C is compact, finitely many of these sets cover C. Thus (7.7) follows from the fact that

$$\begin{aligned}
& t \sum_j P\left\{\max_k \xi_{nj}I_k < t,\ \sum_k \xi_{nj}I_k \geqq t\right\} \\
& \leqq \sum_j E\left[\sum_k \xi_{nj}I_k;\ \max_k \xi_{nj}I_k < t,\ \sum_k \xi_{nj}I_k \geqq t\right] \\
& \leqq \sum_j \sum_h E\left[\xi_{nj}I_h;\ \xi_{nj}I_h < t,\ \sum_k \xi_{nj}I_k \geqq t\right] \\
& = \sum_j \sum_h \left\{E[\xi_{nj}I_h;\ \xi_{nj}I_h < t] - E\left[\xi_{nj}I_h;\ \sum_k \xi_{nj}I_k < t\right]\right\} \\
& = \sum_h \sum_j E[\xi_{nj}I_h;\ \xi_{nj}I_h < t] - \sum_j E[\xi_{nj}I;\ \xi_{nj}I < t] \\
& \to \sum_h \left\{\alpha I_h + \int_0^t x\gamma(dx \times I_h)\right\} - \left\{\alpha I + \int_0^t x\gamma(dx \times I)\right\} = 0\,. \qquad \square
\end{aligned}$$

Let us write $IA(\alpha, \gamma)$ for the class of random measures ξ satisfying (7.2); for point processes we replace $IA(0, \gamma)$ by $IA(\gamma)$. By Table 1, a point process ξ is POISSONian iff $\xi \in IA(\gamma)$ for some γ which is confined to $\{1\} \times \mathfrak{S}$, or equivalently iff $\xi \in I(\lambda)$ for some λ which is confined to the set of degenerate measures with mass 1. Note that in this case

$$E\xi B = \gamma(\{1\} \times B) = \lambda\{\mu B > 0\} , \qquad B \in \mathscr{B} .$$

Corollary 7.4. *Let ξ be a point process on $\mathfrak{S}$ without fixed atoms. Then ξ is Poissonian iff it is a.s. simple and has independent increments.*

Proof. Suppose that ξ is a.s. simple and has independent increments. Then it follows from Theorem 7.1 that ξ is infinitely divisible. Using Theorem 6.7, it is further seen that $\lambda\{\mu_2^* \neq 0\} = 0$, i.e. that $\gamma([2, \infty) \times \mathfrak{S}) = 0$, and so it follows as above that ξ is POISSONian. Conversely, it may e.g. be seen from Theorem 6.7 that any POISSON process without fixed atoms is a.s. simple. □

Specializing the convergence assertion of Theorem 7.2 to the case of POISSONian limits yields the important

Corollary 7.5. *Let ξ be a Poisson process on $\mathfrak{S}$ with intensity $\lambda \in \mathfrak{M}$ and let $\{\xi_{nj}\}$ be a null-array of point processes on $\mathfrak{S}$. Further suppose that $I \subset \mathscr{B}_\lambda$ is a DC-semiring. Then $\sum_j \xi_{nj} \xrightarrow{d} \xi$ iff*

$$\text{(i)} \quad \sum_j P\{\xi_{nj} I > 0\} \to \lambda I , \qquad I \in \mathscr{I} ,$$

$$\text{(ii)} \quad \sum_j P\{\xi_{nj} B > 1\} \to 0 , \qquad B \in \mathscr{B} .$$

Simplicity and diffuseness

We are now going to derive counterparts for null-arrays of Theorems 4.7 and 4.8. For this purpose, we define regularity of canonical measures λ and of null-arrays $\{\xi_{nj}\}$ as follows. Given any covering class $\mathscr{I} \in \mathscr{B}$ and any $a > 0$, we shall say that λ is *a-regular* w.r.t. $\mathscr{I}$, if for every $I \in \mathscr{I}$ there exists some array $\{I_{mk}\} \subset \mathscr{I}$ of finite covers of I such that

$$\lim_{m\to\infty} \sum_k \lambda\{\mu I_{mk} \geqq a\} = 0 . \tag{7.8}$$

For $\{\xi_{nj}\}$ the definition is the same, except that (7.8) is replaced by the relation

$$\lim_{m\to\infty} \limsup_{n\to\infty} \sum_j \sum_k P\{\xi_{nj} I_{mk} \geqq a\} = 0 .$$

As before, *regularity* without any prefix means a-regularity for every $a > 0$.

Theorem 7.6. *Let $\{\xi_{nj}\}$ be a null-array of point processes on $\mathfrak{S}$ and let $\xi \in I(\lambda)$ where $\mu_2^* = 0$ a.e. λ. Further suppose that $\mathscr{U} \subset \mathscr{B}_\lambda$ is a DC-ring while $\mathscr{I} \subset \mathscr{B}_\lambda$ is a DC-semiring. Then $\sum_j \xi_{nj} \xrightarrow{d} \xi$ iff*

$$\text{(i)} \quad \lim_{n\to\infty} \sum_j P\{\xi_{nj} U > 0\} = \lambda\{\mu U > 0\} , \quad U \in \mathscr{U} ,$$

$$\text{(ii)} \quad \limsup_{n\to\infty} \sum_j P\{\xi_{nj} I > 1\} \leqq \lambda\{\mu I > 1\} , \quad I \in \mathscr{I} ,$$

$$\text{(iii)} \quad \lim_{r\to\infty} \limsup_{n\to\infty} \sum_j P\{\xi_{nj} I > r\} = 0 , \quad I \in \mathscr{I}.$$

Moreover, $\sum_j \xi_{nj} \xrightarrow{d} \xi$ *follows from* (i) *and* (ii) *if* λ *is* 2-*regular w.r.t.* $\mathcal{J}$ *and from* (i) *alone if* $\{\xi_{nj}\}$ *is* 2-*regular w.r.t.* $\mathcal{J}$ *or if*

$$\limsup_{n\to\infty} \sum_j E\xi_{nj}I \leqq \lambda\pi_I < \infty\,, \qquad I \in \mathcal{J}\,.$$

The proof is similar to that of Theorem 4.7, except that we now rely on the following two lemmas, here playing the roles of Theorem 3.3 and Lemma 4.6 respectively.

Lemma 7.7. *Let* $\xi \in I(\lambda)$ *and* $\eta \in I(\varkappa)$ *on* $\mathfrak{S}$, *and suppose that* $\mu_2^* = 0$ *a.e.* λ. *Further suppose that* $\mathcal{U} \subset \mathcal{B}$ *is a DC-ring while* $\mathcal{J} \subset \mathcal{B}$ *is a DC-semiring. Writing* $\varphi: \mu \to \mu^*$, *we then have* $\lambda = \varkappa\varphi^{-1}$ *iff*

$$\lambda\{\mu U > 0\} = \varkappa\{\mu U > 0\}\,, \qquad U \in \mathcal{U}\,. \tag{7.9}$$

Furthermore, $\lambda = \varkappa$ *iff* (7.9) *holds and in addition*

$$\lambda\{\mu I > 1\} \geqq \varkappa\{\mu I > 1\}\,, \qquad I \in \mathcal{J}\,.$$

Proof. For fixed $C \in \mathcal{U}$, write $\psi_C: \mu \to C\mu$, and note that (7.9) implies

$$\begin{aligned}\lambda\psi_C^{-1}\{\mu U > 0\} &= \lambda\{(C\mu)\,U > 0\} = \lambda\{\mu(C \cap U) > 0\} = \varkappa\{\mu(C \cap U) > 0\}\\ &= \varkappa\,\{(C\mu)\,U > 0\} = \varkappa\psi_C^{-1}\{\mu U > 0\}\,, \qquad U \in \mathcal{U}\,.\end{aligned}$$

Since $\lambda\psi_C^{-1}(\mathfrak{N}\backslash\{0\}) = \lambda\{\mu C > 0\} < \infty$, it follows from Theorem 3.3 that

$$\lambda\psi_C^{-1}M = \varkappa\psi_C^{-1}\varphi^{-1}M = \varkappa\varphi^{-1}\psi_C^{-1}M\,, \qquad M \in \mathcal{N} \cap (\mathfrak{N}\backslash\{0\})\,.$$

For arbitrary $f \in \mathcal{F}_c$, we may choose a $C \in \mathcal{U}$ containing the support of f, and conclude that

$$\begin{aligned}&\int (1 - e^{-\mu f})\,\lambda(d\mu) = \int (1 - e^{-(C\mu)f})\,\lambda(d\mu) = \int (1 - e^{-\mu f})\,\lambda\psi_C^{-1}(d\mu)\\ &= \int (1 - e^{-\mu f})\,\varkappa\varphi^{-1}\psi_C^{-1}(d\mu) = \int (1 - e^{-(C\mu)f})\,\varkappa\varphi^{-1}(d\mu) = \int (1 - e^{-\mu f})\,\varkappa\varphi^{-1}(d\mu)\,,\end{aligned}$$

which by Theorem 3.1 and the uniqueness assertion of Theorem 6.1 implies that $\lambda = \varkappa\varphi^{-1}$. This proves the first assertion, and the second assertion then follows by applying the counterpart for canonical measures of Lemma 2.7. □

Lemma 7.8. *Let* $\xi \in I(\alpha, \lambda)$ *and let* $\{\xi_{nj}\}$ *be a null-array of random measures on* $\mathfrak{S}$. *Further suppose that* $\mathcal{U} \subset \mathcal{B}$ *is a DC-ring such that*

$$\limsup_{n\to\infty} \sum_j (1 - Ee^{-t\xi_{nj}U}) \leqq t\alpha U + \lambda(1 - e^{-\pi_U})\,, \qquad U \in \mathcal{U}\,, \tag{7.10}$$

for some fixed $t > 0$. *Then* $\mathcal{B}_\eta \supset \mathcal{B}_\alpha \cap \mathcal{B}_\lambda$ *for any random measure* η *such that* $\sum_j \xi_{nj} \xrightarrow{d} \eta$ *as* $n \to \infty$ *through some subsequence. In the point process case, the assertion remains true for* $t = \infty$, *i.e. with* (7.10) *replaced by*

$$\limsup_{n\to\infty} \sum_j P\{\xi_{nj}U > 0\} \leqq \lambda\{\mu U > 0\}\,, \qquad U \in \mathcal{U}\,.$$

The proof is similar to that of Lemma 4.6.

Theorem 7.9. *Let* $\{\xi_{nj}\}$ *be a null-array of random measures* (*or point processes*) *on* $\mathfrak{S}$, *and let* $\xi \in I(\alpha, \lambda)$ *be such that* $\mu \in \mathfrak{M}_d$ *a.e.* λ (*or let* $\alpha = 0$ *and* $\xi \in I(\lambda)$ *with* $\mu_2^* = 0$ *a.e.* λ, *respectively*). *Further suppose that* $s, t \in R$ *are fixed with*

$0 < s < t$, *and that* $\mathcal{U} \subset \mathcal{B}_\xi$ *is a DC-ring while* $\mathcal{I} \subset \mathcal{B}_\xi$ *is a DC-semiring* (*or covering class*). *Then* $\sum_j \xi_{nj} \xrightarrow{d} \xi$ *iff*

(i) $\lim_{n\to\infty} \sum_j (1 - Ee^{-t\xi_{nj}U}) = t\alpha U + \lambda(1 - e^{-t\pi_U})\,, \quad U \in \mathcal{U}\,,$

(ii) $\limsup_{n\to\infty} \sum_j (1 - Ee^{-s\xi_{nj}I}) \leqq s\alpha I + \lambda(1 - e^{-s\pi_I})\,, \quad I \in \mathcal{I}\,,$

(iii) $\lim_{r\to\infty} \limsup_{n\to\infty} \sum_j P\{\xi_{nj} I > r\} = 0\,, \quad I \in \mathcal{I}\,.$

Moreover, $\sum_j \xi_{nj} \xrightarrow{d} \xi$ *follows from* (i) *alone if* $\{\xi_{nj}\}$ *is regular* (*or* 2-*regular*) *w.r.t.* $\mathcal{I}$ *or if*

$$\limsup_{n\to\infty} \sum_j E\xi_{nj} I \leqq \alpha I + \lambda\pi_I < \infty\,, \qquad I \in \mathcal{I}\,.$$

This is a consequence of the following lemma, here playing the role of Theorem 3.4.

Lemma 7.10. *Let* $\xi \in I(\alpha, \lambda)$ *and* $\eta \in I(\beta, \varkappa)$ (*or let* $\alpha = \beta = 0$ *and* $\xi \in I(\lambda)$, $\eta \in I(\varkappa)$), *and suppose that* $\mu \in \mathfrak{M}_d$ (*or* $\mu_2^* = 0$, *respectively*) *a.e.* λ. *Further suppose that* $\mathcal{U} \subset \mathcal{B}$ *is a DC-ring while* $\mathcal{I} \subset \mathcal{B}$ *is a DC-semiring* (*or covering class*), *and that* $s, t \in R$ *are fixed with* $0 < s < t$. *Then* $\xi \stackrel{d}{=} \eta$ *iff* $\mu \in \mathfrak{M}_d$ (*or* $\mu_2^* = 0$) *a.e.* $\varkappa$ *and*

$$t\alpha U + \lambda(1 - e^{-t\pi_U}) = t\beta U + \varkappa(1 - e^{-t\pi_U})\,, \qquad U \in \mathcal{U}\,, \tag{7.11}$$

and also iff (7.11) *holds and in addition*

$$s\alpha I + \lambda(1 - e^{-s\pi_I}) \geqq s\beta I + \varkappa(1 - e^{-s\pi_I})\,, \qquad I \in \mathcal{I}\,. \tag{7.12}$$

If $E\xi \in \mathfrak{M}$, *then* (7.12) *may be replaced by the condition*

$$\alpha I + \lambda\pi_I \geqq \beta I + \varkappa\pi_I\,, \qquad I \in \mathcal{I}\,.$$

Proof. In the random measure case, suppose that $\mu \in \mathfrak{M}_d$ a.e. $\varkappa$ and that (7.11) holds. Consider an arbitrary $x \in \mathfrak{S}$, and let $U \downarrow \{x\}$ in (7.11). Then $\mu U \downarrow 0$ a.e. λ and a.e. $\varkappa$, so we get by dominated convergence $t\alpha\{x\} = t\beta\{x\}$, proving that $\alpha' \equiv \alpha - \alpha_d = \beta - \beta_d \equiv \beta'$. By (6.2) and (7.11) we thus obtain

$$Ee^{-t(\xi - \alpha')U} = Ee^{-t(\eta - \alpha')U}\,, \quad U \in \mathcal{U}\,, \tag{7.13}$$

and since $\xi - \alpha'$ and $\eta - \alpha'$ are both a.s. diffuse by Theorem 6.7, we may conclude from Theorem 3.4 that $\xi - \alpha' \stackrel{d}{=} \eta - \alpha'$, i.e. that $\xi \stackrel{d}{=} \eta$.

Next suppose that (7.11) and (7.12) hold, and proceed as before to conclude that for any $x \in \mathfrak{S}$

$$e^{-t\alpha\{x\}} = Ee^{-t\eta\{x\}}\,, \tag{7.14}$$

$$e^{-s\alpha\{x\}} \leqq Ee^{-s\eta\{x\}}\,. \tag{7.15}$$

But then $\eta\{x\}$ must be degenerate, since A 3.1 would otherwise yield the contradiction

$$e^{-t\alpha\{x\}} = Ee^{-t\eta\{x\}} > (Ee^{-s\eta\{x\}})^{t/s} \geqq (e^{-s\alpha\{x\}})^{t/s} = e^{-t\alpha\{x\}}\,. \tag{7.16}$$

Thus $\eta\{x\} = \alpha\{x\}$ a.s., $x \in \mathfrak{S}$, by (7.14), and it follows in particular that $\eta - \alpha'$ is a random measure. Since (7.11) and (7.12) are equivalent to (7.13) and the relation

$$Ee^{-s(\xi - \alpha')I} \leqq Ee^{-s(\eta - \alpha')I}\,, \qquad I \in \mathcal{I}\,, \tag{7.17}$$

Proof. From Theorem 9.4 it is seen that if the random measures $\xi_j \in S_1(\omega, \alpha_j, \beta_j)$, $j = 1, 2$, are independent, then $\xi_1 + \xi_2 \in S_1(\omega, \alpha_1 + \alpha_2, \beta_1 + \beta_2)$. Similarly, if $\omega\mathfrak{S} = \infty$ while $\xi_j \in S_\infty(\omega, \alpha_j, \gamma_j)$, $j = 1, 2$, are independent, we get by Theorem 9.4 for any $f \in \mathcal{F}$

$$E\left[e^{-(\xi_1+\xi_2)f} \mid \alpha_1, \alpha_2, \gamma_1, \gamma_2\right] = \prod_{j=1}^{2} E[e^{-\xi_j f} \mid \alpha_j, \gamma_j]$$

$$= \prod_{j=1}^{2} \exp\left\{-\alpha_j \omega f - \int_{R'_+ \times \mathfrak{S}} (1 - e^{-xf(s)})\,(\gamma_j \times \omega)\,(dx\,ds)\right\}$$

$$= \exp\left\{-(\alpha_1 + \alpha_2)\,\omega f - \int_{R'_+ \times \mathfrak{S}} (1 - e^{-xf(s)})\,\big((\gamma_1 + \gamma_2) \times \omega\big)\,(dx\,ds)\right\},$$

and taking conditional expectations here, given $(\alpha_1 + \alpha_2, \gamma_1 + \gamma_2)$, it follows that in this case $\xi_1 + \xi_2 \in S_\infty(\omega, \alpha_1 + \alpha_2, \gamma_1 + \gamma_2)$.

Now suppose that $\xi \in S_1(\omega, \alpha, \beta)$ with infinitely divisible (α, β). For any fixed $n \in N$ we may then choose independent random elements $(\alpha_1, \beta_1) \stackrel{d}{=} \cdots \stackrel{d}{=} (\alpha_n, \beta_n)$ with $(\alpha_1 + \cdots + \alpha_n, \beta_1 + \cdots + \beta_n) \stackrel{d}{=} (\alpha, \beta)$. Letting $\xi_j \in S_1(\omega, \alpha_j, \beta_j)$, $j = 1, \dots,$ n, be independent, it follows as above that $\xi_1 + \cdots + \xi_n \in S_1(\omega, \alpha_1 + \cdots + \alpha_n,$ $\beta_1 + \cdots + \beta_n)$, so $\xi_1 + \cdots + \xi_n \stackrel{d}{=} \xi$, and since n was arbitrary, this proves that ξ is infinitely divisible.

Conversely, suppose that $\xi \in S_1(\omega, \alpha, \beta)$ is infinitely divisible, and for $n \in N$, let $\xi_1 \stackrel{d}{=} \cdots \stackrel{d}{=} \xi_n$ be independent with $\xi_1 + \cdots + \xi_n \stackrel{d}{=} \xi$. Considering the corresponding L-transforms, it is seen that $\xi_1, \dots, \xi_n$ are symmetrically distributed w.r.t. ω, so by Theorem 9.4, $\xi_j \in$ some $S_1(\omega, \alpha_j, \beta_j)$, $j = 1, \dots, n$, where (α_1, β_1) $\stackrel{d}{=} \cdots \stackrel{d}{=} (\alpha_n, \beta_n)$ are independent. But then $\xi \stackrel{d}{=} \xi_1 + \cdots + \xi_n \in S_1(\omega, \alpha_1 + \cdots + \alpha_n,$ $\beta_1 + \cdots + \beta_n)$, so $(\alpha_1 + \cdots + \alpha_n, \beta_1 + \cdots + \beta_n) \stackrel{d}{=} (\alpha, \beta)$, and n being arbitrary, this shows that (α, β) is infinitely divisible, and hence completes the proof for $\omega\mathfrak{S} < \infty$. The proof for $\omega\mathfrak{S} = \infty$ is similar. □

To state the next result, denote by $\mathfrak{h}$ the function $\mathfrak{h}(x) \equiv x$.

Theorem 9.6. *Let $\omega \in \mathfrak{M}_d$ and let $\xi \in I(\Gamma, \Lambda)$ on $\mathfrak{S}$. Then ξ is symmetrically distributed w.r.t. ω iff*

for $0 < \omega\mathfrak{S} < \infty$, there exist some $a \in R_+$ and some measure λ on the set $\{(x, \mu) \in R_+ \times \mathfrak{N}(R'_+)\backslash\{0\} : \mu\mathfrak{h} < \infty\}$ satisfying $\int \mathfrak{g}(x + \mu\mathfrak{h})\,\lambda(dx\,d\mu) < \infty$, such that $\Gamma = a\omega/\omega\mathfrak{S}$ while $\Lambda = \int P\xi_{x,\mu}^{-1}\,\lambda(dx\,d\mu)$ where $\xi_{x,\mu} \in S_1(\omega, x, \mu)$, while

for $\omega\mathfrak{S} = \infty$, there exist some $a \in R_+$, some $m \in \mathfrak{M}(R'_+)$ with $m\mathfrak{g} < \infty$ and some measure λ on the set $\{(x, \mu) \in R_+ \times \mathfrak{M}(R'_+)\backslash\{0\} : \mu\mathfrak{g} < \infty\}$ satisfying $\int \mathfrak{g}(x + \mu\mathfrak{g})\,\lambda(dx\,d\mu) < \infty$, such that $\Gamma = a\omega$ while

$$\Lambda = \iint \delta_{x\delta_s} m(dx)\,\omega(ds) + \int P\eta_{x,\mu}^{-1}\lambda(dx\,d\mu)\,, \tag{9.14}$$

where $\eta_{x,\mu} \in S_\infty(\omega, x, \mu)$.

In this case the pair (a, λ) or the triple (a, m, λ) is unique and it is further determined by the property that $\xi \in S_1(\omega, \alpha, \beta)$ for some $(\alpha, \beta) \in I\big((a, 0), \lambda\big)$ or that $\xi \in S_\infty(\omega, \alpha, \gamma)$ for some $(\alpha, \gamma) \in I\big((a, m), \lambda\big)$ respectively.

Proof. Suppose that $0 < \omega\mathfrak{S} < \infty$ and let $\xi \in S_1(\omega, \alpha, \beta)$ be infinitely divisible. By Lemma 9.5 and Theorem 6.1, there exist some $a \in R_+$ and some measure λ on $R_+ \times \mathfrak{N}(R'_+)\backslash\{0\}$ satisfying

$$\int \mathfrak{g}(x)\,\lambda(dx\,d\mu) < \infty\,; \quad \int \mathfrak{g}(\mu[s,t])\,\lambda(dx\,d\mu) < \infty\,, \quad 0 < s < t < \infty\,, \tag{9.15}$$

such that $(\alpha, \beta) \in I((a, 0), \lambda)$, i.e.

$$-\log Ee^{-\alpha r-\beta g} = ar + \int (1 - e^{-xr-\mu g})\,\lambda(dx\,d\mu)\,, \quad r \in R_+\,, \quad g \in \mathcal{F}(R'_+)\,, \tag{9.16}$$

and by Theorems 6.1 and 9.4, a and λ are uniquely determined by $P\xi^{-1}$. Taking $r = 0$ and $g = t\mathfrak{h}$ in (9.16) and letting $t \to 0$, it is further seen that $\beta\mathfrak{h} < \infty$ a.s. iff $\lambda\{(x, \mu): \mu\mathfrak{h} = \infty\} = 0$ and $\int \mathfrak{g}(\mu\mathfrak{h})\,\lambda(dx\,d\mu) < \infty$. Since the latter condition contains the second part of (9.15), and since moreover $\mathfrak{g}(x) \vee \mathfrak{g}(\mu\mathfrak{h}) \leqq \mathfrak{g}(x + \mu\mathfrak{h}) \leqq \mathfrak{g}(x) + \mathfrak{g}(\mu\mathfrak{h})$, the stated conditions on a and λ are both necessary and sufficient. In the case $\omega S = \infty$, it follows by similar arguments that $\xi \in S_\infty(\omega, \alpha, \gamma)$ for some $(\alpha, \gamma) \in I((a, m), \lambda)$, i.e. that

$$-\log Ee^{-\alpha r-\gamma g} = ar + mg + \int (1 - e^{-xr-\mu g})\,\lambda(dx\,d\mu)\,, \quad r \in R_+\,, \quad g \in \mathcal{F}(R'_+)\,, \tag{9.17}$$

where a, m and λ are unique with the stated properties.

Still assuming $\omega\mathfrak{S} = \infty$, let $f \in \mathcal{F}$, put $g(x) \equiv \omega(1 - e^{-xf})$ and conclude from (9.17) that

$$\begin{aligned} Ee^{-\xi f} = EE\,[e^{-\xi f} \mid \alpha, \gamma] = Ee^{-\alpha\omega f-\gamma g} \\ = \exp\{-a\omega f - mg - \int (1 - e^{-x\omega f-\mu g})\,\lambda(dx\,d\mu)\}\,. \end{aligned}$$

Since $e^{-x\omega f-\mu g} \equiv Ee^{-\eta_{x,\mu}f}$, it follows that

$$-\log Ee^{-\xi f} = a\omega f + \int (1 - e^{-xf(s)})\,m(dx)\,\omega(ds) + \int E\,(1 - e^{-\eta_{x,\mu}f})\,\lambda(dx\,d\mu)\,,$$

and so it is seen by identification that Γ and Λ have the stated forms.

In the case $0 < \omega\mathfrak{S} < \infty$, assume for simplicity of writing that $\omega\mathfrak{S} = 1$, and define $g(x) \equiv -\log \omega e^{-xf}$. We may then conclude from (9.16) that

$$\begin{aligned} Ee^{-\xi f} &= EE\left[\exp\left(-\alpha\omega f - \sum_j \beta_j f(\tau_j)\right) \Big| (\alpha, \beta)\right] = Ee^{-\alpha\omega f} \prod_j E\,[e^{-\beta_j f(\tau_j)} \mid \beta_j] \\ &= Ee^{-\alpha\omega f} \prod_j \exp(-g(\beta_j)) = Ee^{-\alpha\omega f} \exp\left(-\sum_j g(\beta_j)\right) = Ee^{-\alpha\omega f-\beta g} \\ &= \exp\{-a\omega f - \int (1 - e^{-x\omega f-\mu g})\,\lambda(dx\,d\mu)\}\,. \end{aligned}$$

These calculations show in particular that $e^{-x\omega f-\mu g} \equiv Ee^{-\xi_{x,\mu}f}$, so we get

$$-\log Ee^{-\xi f} = a\omega f + \int E(1 - e^{-\xi_{x,\mu}f})\,\lambda(dx\,d\mu)\,,$$

proving that Γ and Λ have again the stated forms. □

Notes. Theorem 9.1, being new for $t < \infty$, was proved for $t = \infty$ in [31], and independently (for LEBESGUE intervals) by DAVIDSON [10], while Corollary 9.2 was given in [31] and independently in KERSTAN, MATTHES and MECKE [44][1]), p. 80, after a special case had been treated by NAWROTZKI [62]. (The direct assertion is of course classical in the POISSONian case.) For Corollary 9.3 and some further results in that direction, we refer to [36]. Theorem 9.4 was proved for unbounded ω by BÜHLMANN [8] (in the LEBESGUE interval case; the general case was treated in [30]), and for bounded ω in [32]. (Cf. [37] for a different proof, and also [33] for extensions to random processes and for the connection with thinning.) Finally, special cases of Lemma 9.5 and Theorem 9.6 were given by KERSTAN, MATTHES and MECKE [44], pp. 76—77, while the general results are essentially contained in [39].

[1]) See also DAVIDSON [10], p. 39.

Exercises

9.1. Let $\omega \in \mathfrak{M}_d$. Show that a random measure ξ on $\mathfrak{S}$ is symmetrically distributed w.r.t. ω iff the random variables $\xi B_1, \dots, \xi B_k$ are interchangeable for any $k \in N$ and disjoint $B_1, \dots, B_k \in \mathcal{B}$ with $\omega B_1 = \dots = \omega B_k$. (Cf. [32]. Hint: Approximate $\omega B_1, \dots, \omega B_k$ in the symmetry definition by binary rationals or scrutinize the proof of Theorem 9.4.)

9.2. Let $\omega \in \mathfrak{M}_d$ and suppose that ξ is a β-compound of η, where we assume that $\beta \overset{d}{\neq} 0$. Show that ξ and η are simultaneously symmetrically distributed w.r.t. ω. Prove also the corresponding result for the random measures defined by (8.7) and (8.8). (Hint: Use the corresponding uniqueness results, being obtained by analytic continuation.)

9.3. For fixed $k, n \in N$ with $2 \leqq k \leqq n$, let $\omega = \sum_{j=1}^{n} \delta_j$ and $\xi = \sum_{j \in K} \delta_j$, where K is a randomly chosen subset of $\{1, \dots, n\}$ containing k elements. Show that ξ is (in the obvious sense) symmetrically distributed w.r.t. ω and still not a mixed sample process. (Hint: Note that ξ is simple.) Thus the assumption $\omega \in \mathfrak{M}_d$ is essential in Theorem 9.1.

9.4. Starting from Corollary 9.3, state and prove a corresponding criterion for the convergence $\xi_n \overset{d}{\to} \xi$, where ξ is a fixed Poisson or sample process with diffuse intensity, the latter being unbounded in the Poissonian case. (Cf. [36].)

9.5. Let ξ be symmetrically distributed w.r.t. some $\omega \in \mathfrak{M}_d$, suppose that $C \in \mathcal{B}$ with $0 < \omega C < \omega \mathfrak{S}$, and let α and β be a.s. defined by $C\xi \in S_1(C\omega, \alpha, \beta)$. Show that β cannot be non-random unless $\xi = \alpha\omega/\omega C$ a.s. (Hint: Use the thinning interpretation of β.)

9.6. Let $\omega \in \mathfrak{M}_d$ be fixed and let ξ be symmetrically distributed w.r.t. ω. Show that $P\xi^{-1}$ is uniquely determined by $P(C\xi)^{-1}$ for any fixed $C \in \mathcal{B}$ with $\omega C > 0$. (Cf. [33]. Hint: Use an extended version of Corollary 3.2.)

9.7. Use Theorem 9.4 to extend Corollary 9.2 to the case of arbitrary symmetrically distributed random measures. (Cf. Westcott [77].)

9.8. Show that the conclusion of Lemma 9.5 is false in general when $\xi \in$ some $S_\infty(\omega, \alpha, \gamma)$ with $\omega\mathfrak{S} < \infty$. (Hint: Consider Exercise 8.6.)

9.9. Restate the results of Theorem 9.6 in terms of cluster representations. (The use of such explicit representations lead in [39] to substantial simplifications of proofs.)

9.10. Specialize Theorems 9.4 and 9.6 to the case of point processes.

10. Palm distributions

Definitions and basic properties

For any random measure ξ on $\mathfrak{S}$, we define the Campbell *measure* $(P\xi^{-1})^1$ on $\mathfrak{S} \times \mathfrak{M}$ by

$$(P\xi^{-1})^1 \, (B \times M) = E[\xi B; \xi \in M], \qquad B \in \mathcal{B}, \quad M \in \mathcal{M}.$$

(Note that this is a refinement of the intensity or first moment measure of ξ, hence the superscript 1. Campbell measures of higher order may be defined analogeously.) If $E\xi$ is σ-finite, which holds e.g. when $E\xi \in \mathfrak{M}$, we may define the corresponding Radon-Nikodym derivatives $P_s M$ by

$$P_s M = \frac{(P\xi^{-1})^1 \, (ds \times M)}{(P\xi^{-1})^1 \, (ds \times \mathfrak{M})} = \frac{E[\xi(ds); \xi \in M]}{E\xi(ds)}, \quad s \in \mathfrak{S} \text{ a.e. } E\xi, \quad M \in \mathcal{M}. \tag{10.2}$$

More generally, if $f \in \mathcal{F}$ is such that $Ef\xi$ is σ-finite, we may define

$$P_s M = \frac{E[f\xi(ds);\, \xi \in M]}{Ef\xi(ds)}, \qquad s \in \mathfrak{S} \text{ a.e. } Ef\xi, \qquad M \in \mathcal{M}, \quad (10.3)$$

without ambiguity, since the expressions (10.3) obtained for two different functions f_1 and f_2 coincide a.e. $E\xi$ on the support of $f_1 \wedge f_2$. Note that, for fixed $M \in \mathcal{M}$, $P_s M$ is by definition an $\mathcal{S}$-measurable function of s. Since $\mathfrak{M}$ is Polish by A 7.7, it is further possible to choose versions of $P_s M$ such that P_s becomes a probability measure on $\mathfrak{M}$ for each $s \in \mathcal{S}$. (Note the similarity with the case of conditional distributions.) We shall always assume such versions to be chosen, and the P_s will then be referred to as the PALM *distributions* of ξ. For convenience, we introduce a family $\{\xi_s\}$ of random measures on $\mathfrak{S}$ such that $P\xi_s^{-1} = P_s$, $s \in \mathfrak{S}$.

Lemma 10.1. *Let ξ be a random measure on $\mathfrak{S}$ such that $E\xi$ is σ-finite, and let $f \in \mathcal{F}(\mathfrak{S} \times \mathfrak{M})$. Then*

$$Ef(s, \xi_s) = \frac{E[f(s, \xi)\, \xi(ds)]}{E\xi(ds)}, \qquad s \in \mathfrak{S} \text{ a.e. } E\xi .$$

Proof. By the definition of RADON-NIKODYM derivatives, we have to show that $Ef(s, \xi_s)$ is $\mathcal{S}$-measurable and satisfies

$$\int_B Ef(s, \xi_s)\, E\xi(ds) = E \int_B f(s, \xi)\, \xi(ds), \qquad B \in \mathcal{S} . \quad (10.4)$$

Now (10.4) holds by definition of ξ_s for indicators $f = 1_{A \times M}$, where $A \in \mathcal{S}$ with $E\xi A < \infty$ while $M \in \mathcal{M}$, and by means of A 2.2 and monotone convergence, it may easily be extended to arbitrary indicators over $\mathcal{S} \times \mathcal{M}$. But then it is true for simple functions over $\mathcal{S} \times \mathcal{M}$, and hence by monotone convergence for arbitrary f. □

In the point process case, it is sometimes natural to consider $\xi_s - \delta_s$ in place of ξ_s. This is justified by

Lemma 10.2. *Let ξ be a point process on $\mathfrak{S}$ such that $E\xi$ is σ-finite. Then $\xi_s - \delta_s$ is a point process on $\mathfrak{S}$ for $s \in \mathfrak{S}$ a.e. $E\xi$, and moreover*

$$Ef(s, \xi_s - \delta_s) = \frac{Ef(s, \xi - \delta_s)\, \xi(ds)}{E\xi(ds)}, \qquad s \in \mathfrak{S} \text{ a.e. } E\xi , \qquad f \in \mathcal{F}(\mathfrak{S} \times \mathfrak{N}). \quad (10.5)$$

Proof. First note that

$$\{(s, \mu) \in \mathfrak{S} \times \mathfrak{M} : \mu - \delta_s \in \mathfrak{N}\} = \{(s, \mu) \in \mathfrak{S} \times \mathfrak{N} : \mu\{s\} \geqq 1\} ,$$

where the right-hand side belongs to $\mathcal{S} \times \mathcal{N}$, since for any $B \in \mathcal{B}$ and any null-array $\{B_{nj}\} \subset \mathcal{B}$ of partitions of B,

$$\{(s, \mu) \in B \times \mathfrak{N} : \mu\{s\} \geqq 1\} = \bigcap_n \bigcup_j (B_{nj} \times \{\mu \in \mathfrak{N} : \mu \bar{B}_{nj} \geqq 1\}) \in \mathcal{B} \times \mathcal{N} ,$$

and since $\mathfrak{S}$ is σ-compact. Thus we get by (10.4) for any $B \in \mathcal{B}$

$$\int_B P\{\xi_s - \delta_s \in \mathfrak{N}\}\, E\xi(ds) = E \int_B 1_N(\xi - \delta_s)\, \xi(ds) = E \int_B \xi(ds) = E\xi B ,$$

so $\int_B (1 - P\{\xi_s - \delta_s \in \mathfrak{N}\})\, E\xi(ds) = 0$ and hence

$$P\{\xi_s - \delta_s \in \mathfrak{N}\} = 1 , \qquad s \in \mathfrak{S} \text{ a.e. } E\xi .$$

We may now apply Lemma 10.1 to obtain (10.5). □

To avoid the above arbitrariness in the definition of P_s, we may mix w.r.t. some bounded measure which is absolutely continuous w.r.t. $E\xi$. We thus introduce for fixed $f \in \mathcal{F}$ with $0 < E\xi f < \infty$ a random measure ξ_f on $\mathfrak{S}$ with distribution

$$P\{\xi_f \in M\} = \frac{1}{E\xi f} \int_{\mathfrak{S}} P\{\xi_s \in M\}\, f(s)\, E\xi(ds) = \frac{E[\xi f;\, \xi \in M]}{E\xi f}, \qquad M \in \mathcal{M}.$$

In the point process case, we further introduce a point process ξ'_f on $\mathfrak{S}$ satisfying

$$P\{\xi'_f \in M\} = \frac{1}{E\xi f} \int_{\mathfrak{S}} P\{\xi_s - \delta_s \in M\}\, f(s)\, E\xi(ds) = \frac{1}{E\xi f} E \int_{\mathfrak{S}} 1_M(\xi - \delta_s)\, f(s)\, \xi(ds), \qquad M \in \mathcal{N}.$$

In analogy with Lemma 10.1, we obtain the formulae

$$Eg(\xi_f) = \frac{1}{E\xi f} \int_{\mathfrak{S}} Eg(\xi_s)\, f(s)\, E\xi(ds) = \frac{Eg(\xi)\, \xi f}{E\xi f}, \tag{10.6}$$

$$Eg(\xi'_f) = \frac{1}{E\xi f} \int_{\mathfrak{S}} Eg(\xi_s - \delta_s)\, f(s)\, E\xi(ds) = \frac{1}{E\xi f} E \int_{\mathfrak{S}} g(\xi - \delta_s)\, f(s)\, \xi(ds), \tag{10.7}$$

being valid for arbitrary $g \in \mathcal{F}(\mathfrak{M})$ or $\mathcal{F}(\mathfrak{N})$ respectively.

Uniqueness and continuity

Lemma 10.3. *Let ξ be a random measure on $\mathfrak{S}$ with known intensity $E\xi \in \mathfrak{M}$. Then $P\xi^{-1}$ is uniquely determined by $P\xi_f^{-1}$ for all $f \in \mathcal{F}_c$ with $E\xi f > 0$. In the point process case, this statement remains true with ξ_f replaced by ξ'_f.*

Proof. Suppose that $E\xi = E\eta$ and that $P\xi_f^{-1} = P\eta_f^{-1}$ for all $f \in \mathcal{F}_c$ with $E\xi f > 0$. Letting $f \in \mathcal{F}_c$ be fixed and putting

$$g(\mu) = \begin{cases} (1 - e^{-\mu f})/\mu f, & \mu f > 0, \\ 1, & \mu f = 0, \end{cases}$$

we get by (10.6)

$$1 - L_\xi(f) = E\xi f g(\xi) = E\eta f g(\eta) = 1 - L_\eta(f), \tag{10.8}$$

and it follows by Theorem 3.1 that $\xi \overset{d}{=} \eta$. In the point process case, we may conclude from (10.7) with f replaced by $h = fe^{-tf}$ and with $g(\mu) = e^{-t\mu f}$ that, for $t \in R_+$,

$$E\xi f e^{-t\xi f} = E \int g(\xi - \delta_s)\, h(s)\, \xi(ds) = E \int g(\eta - \delta_s)\, h(s)\, \eta(ds) = E\eta f e^{-t\eta f},$$

and hence that

$$1 - L_\xi(f) = E \int_0^1 \xi f e^{-t\xi f}\, dt = E \int_0^1 \eta f e^{-t\eta f}\, dt = 1 - L_\eta(f). \qquad \square$$

The uniqueness may be strengthened to continuity in the following sense.

Theorem 10.4. *Let $\xi, \xi_1, \xi_2, \ldots$ be random measures on $\mathfrak{S}$, and suppose that $E\xi \in \mathfrak{M}\setminus\{0\}$. Then any two of the following conditions imply the third one.*

(i) $E\xi_n \xrightarrow{v} E\xi$,

(ii) $\xi_n \xrightarrow{d} \xi$,

(iii) $(\xi_n)_f \xrightarrow{d} \xi_f$ *for all* $f \in \mathcal{F}_c$ *with* $E\xi f > 0$.

Proof. Suppose that (i) and (ii) hold, let $f \in \mathcal{F}_c$ with $E\xi f > 0$, and let $g \in \mathcal{F}(\mathfrak{M})$ be bounded and continuous. Then the mapping $\mu \to g(\mu)\,\mu f$ is continuous by A 7.3, so by A 4.2

$$g(\xi_n)\,\xi_n f \xrightarrow{d} g(\xi)\,\xi f\,. \tag{10.9}$$

Furthermore, it follows by A 4.3 that the variables $\xi_n f$ and hence also $g(\xi_n)\,\xi_n f$ are (asymptotically) uniformly integrable, so (10.9) yields

$$Eg(\xi_n)\,\xi_n f \to Eg(\xi)\,\xi f\,, \tag{10.10}$$

which by (10.6) implies

$$Eg\big((\xi_n)_f\big) \to Eg(\xi_f)\,. \tag{10.11}$$

Since g was arbitrary, we thus obtain $(\xi_n)_f \xrightarrow{d} \xi_f$, and (iii) follows.

Next suppose that (ii) and (iii) hold, and define $g(\mu) \equiv (1 + \mu f)^{-1}$ for arbitrary $f \in \mathcal{F}_c$ with $E\xi f > 0$. Then $g(\mu)$ and $g(\mu)\,\mu f$ are both bounded and continuous, so (10.10) and (10.11) are satisfied, and by (10.6) this entails $E\xi_n f \to E\xi f$. Taking differences, it is seen that this result remains true when $E\xi f = 0$, and so (i) is proved.

Finally suppose that (i) and (iii) hold, and let $f \in \mathcal{F}_c$ be arbitrary with $E\xi f > 0$. Then

$$\lim_{t\to\infty} \limsup_{n\to\infty} E[\xi_n f;\, \xi_n f \geqq t] \leqq \lim_{t\to\infty} E[\xi f;\, \xi f \geqq t] = 0$$

by A 4.1, so the random variables $\xi_n f$ are uniformly integrable, and by (i) this is also true when $E\xi f = 0$. Now $\{\xi_n\}$ is relatively compact by (i) and Lemma 4.5, so any sequence $N' \subset N$ must contain a subsequence N'' such that $\xi_n \xrightarrow{d}$ some η ($n \in N''$). By A 4.3 it follows that $E\xi_n \xrightarrow{v} E\eta$ ($n \in N''$), and so we may conclude as above that $(\xi_n)_f \xrightarrow{d} \eta_f$ for all $f \in \mathcal{F}_c$ with $E\eta f > 0$. Comparing this with (i) and (iii) and using Lemma 10.3, we obtain $\xi \overset{d}{=} \eta$, and so we may apply A 1.2 to complete the proof. □

Theorem 10.5. *In the point process case, the conclusion of Theorem 10.4 remains true with $(\xi_n)_f$ and ξ_f replaced by $(\xi_n)'_f$ and ξ'_f respectively.*

Proof. Consider any $f \in \mathcal{F}_c$ with $E\xi f > 0$ and any bounded continuous $g \in \mathcal{F}(\mathfrak{N})$, and let us first show that the function

$$h(\mu) = \int g(\mu - \delta_s)\, f(s)\, \mu(ds)\,, \qquad \mu \in \mathfrak{N}\,, \tag{10.12}$$

is vaguely continuous. For this purpose, let $\mu, \mu_1, \mu_2, \ldots \in \mathfrak{N}$ with $\mu_n \xrightarrow{v} \mu$ and consider arbitrary atom positions $s, s_1, s_2, \ldots$ of $\mu, \mu_1, \mu_2, \ldots$ satisfying $s_n \to s$. Then $\mu_n - \delta_{s_n} \xrightarrow{v} \mu - \delta_s$, so we get by continuity $g(\mu_n - \delta_{s_n})\, f(s_n) \to g(\mu - \delta_s)\, f(s)$. The functions g and f being bounded, the latter with bounded support, it follows by A 7.3 that $h(\mu_n) \to h(\mu)$, as desired. Assuming (i) and (ii) to hold, we may now argue as in the preceding proof to conclude from A 4.2, A 4.3 and (10.7) that $Eg((\xi_n)'_f) \to Eg(\xi'_f)$, and since g was arbitrary, this proves (iii).

Next suppose that (ii) and (iii) are fulfilled, and define

$$g(\mu) = (1 + \mu f + \|f\|)^{-1}, \qquad \mu \in \mathfrak{N}.$$

This g being bounded and continuous, we get by (iii)

$$Eg((\xi_n)'_f) \to Eg(\xi'_f) > 0. \tag{10.13}$$

Moreover, the function h in (10.12) was shown above to be continuous, and in the present case it is even bounded since

$$g(\mu) = \int \frac{f(s)\,\mu(ds)}{1 + \mu f - f(s) + \|f\|} \leqq \int \frac{f(s)\,\mu(ds)}{1 + \mu f} = \frac{\mu f}{1 + \mu f} \leqq 1,$$

so we get $Eh(\xi_n) \to Eh(\xi)$ by (ii), and it follows by (10.7) and (10.13) that $E\xi_n f \to E\xi f$. Taking differences, it is seen that this result remains true when $E\xi f = 0$, and so (i) is proved.

Finally suppose that (i) and (ii) hold, and let $f \in \mathcal{F}_c$ be arbitrary with $E\xi f > 0$. Then

$$E[\xi_n f; \xi_n f \geqq t] \leqq E \int_{\mathfrak{S}} 1_{\{\mu: \mu f \geqq t - \|f\|\}}(\xi_n - \delta_s)\, f(s)\, \xi(ds)$$

$$= (E\xi_n f)\, P\{(\xi_n)'_f \geqq t - \|f\|\}, \qquad n \in N, \qquad t \in R_+,$$

so we get

$$\lim_{t\to\infty} \limsup_{n\to\infty} E[\xi_n f; \xi_n f \geqq t] \leqq (E\xi f) \lim_{t\to\infty} P\{\xi'_f \geqq t - \|f\|\} = 0,$$

which proves that the variables $\xi_n f$ are uniformly integrable. We may thus complete the proof as in case of Theorem 10.4. □

Infinite divisibility

Let $\xi \in I(\alpha, \lambda)$ be such that $E\xi$ is σ-finite, and define the CAMPBELL measure Λ on $\mathfrak{S} \times \mathfrak{M}$ by $\Lambda = \alpha\delta_0 + \lambda^1$ (cf. (10.1)), or more explicitly

$$\Lambda(B \times M) = \alpha B \delta_0 M + \int_M \mu B \lambda(d\mu), \qquad B \in \mathcal{B}, \qquad M \in \mathcal{M}.$$

Since $\Lambda(B \times \mathfrak{M}) = E\xi B$, $B \in \mathcal{B}$, (cf. Exercise 6.6), we may proceed as in (10.2) to define a family $\{\Lambda_s\}$ of distributions on $(\mathfrak{M}, \mathcal{M})$ by

$$\Lambda_s M = \frac{\Lambda(ds \times M)}{\Lambda(ds \times \mathfrak{M})} = \frac{\alpha(ds)\,\delta_0 M + \int_M \mu(ds)\,\lambda(d\mu)}{\alpha(ds) + \int_{\mathfrak{M}} \mu(ds)\,\lambda(d\mu)}, \qquad s \in \mathfrak{S} \text{ a.e. } E\xi.$$

Introducing for convenience the random measures $\tilde{\xi}_s$ on $\mathfrak{S}$ possessing these distributions, we get as in Lemma 10.1 for any $f \in \mathcal{F}(\mathfrak{S} \times \mathfrak{M})$

$$Ef(s, \tilde{\xi}_s) = \frac{\alpha(ds)\, f(s, 0) + \int_{\mathfrak{M}} f(s, \mu)\,\mu(ds)\,\lambda(d\mu)}{\alpha(ds) + \int_{\mathfrak{M}} \mu(ds)\,\lambda(d\mu)}, \qquad s \in \mathfrak{S} \text{ a.e. } E\xi. \tag{10.14}$$

For any $f \in \mathcal{F}$ we further obtain

$$|(L_\zeta L_{\eta_s} - L_\xi L_{\eta_s})(f)| \leqq |L_\zeta(f) - L_\xi(f)|$$
$$= \left|\frac{E[e^{-\xi f};\ \xi_\varepsilon^* B = 0] - (Ee^{-\xi f})\, P\{\xi_\varepsilon^* B = 0\}}{P\{\xi_\varepsilon^* B = 0\}}\right|$$
$$= \left|\frac{E[e^{-\xi f};\ \xi_\varepsilon^* B > 0] - (Ee^{-\xi f})\, P\{\xi_\varepsilon^* B > 0\}}{P\{\xi_\varepsilon^* B = 0\}}\right| \leqq \frac{P\{\xi_\varepsilon^* B > 0\}}{P\{\xi_\varepsilon^* B = 0\}}.$$

Combining these two estimates, we get for $s \in B$ a.e. $E\xi$

$$|L_{\xi_s} - L_\xi L_{\eta_s}| \leqq \frac{1}{P\{\xi_\varepsilon^* B = 0\}}\left(\frac{E\xi'_\varepsilon(ds)}{E\xi(ds)} + P\{\xi_\varepsilon^* B > 0\}\right). \tag{11.7}$$

If B is restricted to some countable class $\mathcal{J} \subset \mathcal{B}$ and ε to the positive rationals Q'_+, there exists some fixed set $C \in \mathcal{S}$ with $E\xi C = 0$ such that (11.7) holds whenever $P\{\xi_\varepsilon^* B = 0\} > 0$ and $s \in B \backslash C$. Here we choose $\mathcal{J}$ such that $E\xi B < \infty$, $B \in \mathcal{J}$, and such that moreover, given any $s \in \mathfrak{S}$, there exist some $B_1, B_2, \ldots \in \mathcal{J}$ with $B_n \downarrow \{s\}$ and $|B_n| \downarrow 0$.

Since $\xi'_\varepsilon \xrightarrow{v} 0$ a.s. as $\varepsilon \to 0$, we may clearly choose versions of the derivatives $E\xi'(ds)/E\xi(ds)$ which are non-decreasing in $\varepsilon \in Q'_+$, and in this case we obtain by dominated convergence for any $B \in \mathcal{B}$ with $E\xi B < \infty$

$$0 = E \lim_{\varepsilon\to 0} \xi'_\varepsilon B = \lim_{\varepsilon\to 0} E\xi'_\varepsilon B = \lim_{\varepsilon\to 0} \int_B \frac{E\xi'_\varepsilon(ds)}{E\xi(ds)}\, E\xi(ds)$$
$$= \int_B \lim_{\varepsilon\to 0} \frac{E\xi'_\varepsilon(ds)}{E\xi(ds)}\, E\xi(ds) \qquad (\varepsilon \in Q'_+),$$

which proves that $E\xi'(ds)/E\xi(ds) \to 0$ $(\varepsilon \in Q'_+)$ for all $s \notin$ some $C' \in \mathcal{S}$ with $E\xi C' = 0$. For any fixed $n \in N$ and $s \in \mathfrak{S}\backslash(C \cup C')$, we may now choose $\varepsilon \in Q'_+$ so small that this derivative is less than $(4n)^{-1}$, and then $B \in \mathcal{J}$ so small that $P\{\xi_\varepsilon^* B > 0\} < (4n)^{-1}$, the latter choice being possible since the point process ξ_ε^* has no fixed atoms. Writing ζ_n for the corresponding random measure η_s, it is then seen from (11.7) that $|L_{\xi_s} - L_\xi L_{\zeta_n}| < n^{-1}$. Hence $L_\xi L_{\zeta_n} \to L_{\xi_s}$ as $n \to \infty$, and so it follows easily by Lemma 4.5 that $L_\xi L_\zeta = L_{\xi_s}$ for some random measure ζ, i.e. that $P\xi^{-1} \mid P\xi_s^{-1}$. This being true for any $s \notin (C \cup C')$, we may thus conclude from Theorem 11.2 that ξ is infinitely divisible. For discrete ξ, the same argument applies with $\varepsilon = 0$ throughout. □

Mixed Poisson and sample processes

Theorem 11.5. *Let ξ be a point process on $\mathfrak{S}$ such that $E\xi$ is σ-finite. Then $P(\xi_s - \delta_s)^{-1}$ is independent of $s \in \mathfrak{S}$ a.e. $E\xi$ iff $E\xi \in \mathfrak{M}$ and $\xi \in M(E\xi, \varphi)$ for some φ. In this case $\xi_s - \delta_s \in M(E\xi, -\varphi')$, $s \in \mathfrak{S}$ a.e. $E\xi$.*

Here and below we shall need the following lemma of some independent interest.

Lemma 11.6. *Let ξ be a random measure on $\mathfrak{S}$ such that $E\xi$ is σ-finite, and let $C \in \mathcal{B}$ be fixed. Further suppose that τ is a random element in $\mathfrak{S}$ satisfying*

$P\{\tau \in \cdot \mid \xi\} = C\xi/\xi C$ *a.s. on* $\{\xi C > 0\}$. *Then*

$$P\{(\xi, \tau) \in \cdot \mid \xi C = x,\ \tau = s\} = P\{(\xi_s, s) \in \cdot \mid \xi_s C = x\}\,,$$
$$(x,s) \in R_+ \times C \ \text{a.e.}\ E[\xi(ds);\ \xi C \in dx]\,.$$

Proof. For $(x,s) \in R_+ \times C$ and for any fixed $f \in \mathcal{F}(\mathfrak{M} \times \mathfrak{S})$ we get by Lemma 10.1

$$E[f(\xi, \tau);\ \xi C \in dx,\ \tau \in ds] = E[f(\xi, s)\, P\{\tau \in ds \mid \xi\};\ \xi C \in dx]$$
$$= E[f(\xi, s)\, \xi(ds)/x;\ \xi C \in dx] = x^{-1}\, E[f(\xi_s, s);\ \xi_s C \in dx]\, E\xi(ds)\,,$$

and in particular

$$P\{\xi C \in dx,\ \tau \in ds\} = x^{-1}\, P\{\xi_s C \in dx\}\, E\xi(ds)\,,$$

so applying the chain rule for RADON-NIKODYM derivatives, we obtain a.e.

$$E[f(\xi, \tau) \mid \xi C = x,\ \tau = s] = \frac{E[f(\xi, \tau);\ \xi C \in dx,\ \tau \in ds]}{P\{\xi C \in dx,\ \tau \in ds\}} = \frac{E[f(\xi_s, s);\ \xi_s C \in dx]}{P\{\xi_s C \in dx\}}$$
$$= E[f(\xi_s, s) \mid \xi_s C = x]\,.$$

Since by Theorem 3.1 countably many expectations determine a distribution on $\mathfrak{M} \times \mathfrak{S}$, this completes the proof. □

Proof of Theorem 11.5. Assuming that $\xi \in M(E\xi, \varphi)$, we get by (9.1) for any $f \in \mathcal{F}$ and $B \in \mathcal{B}$

$$E\xi B e^{-\xi f} = -\frac{d}{dt} L_\xi(f + t1_B)\Big|_{t=0} = -\frac{d}{dt}\varphi\left(E\xi(1 - e^{-f - t1_B})\right)\Big|_{t=0}$$
$$= -\varphi'\left(E\xi(1 - e^{-f})\right) \int_B e^{-f(s)}\, E\xi(ds)\,,$$

so

$$E\xi(ds)\, e^{-\xi f} = -\varphi'\left(E\xi(1 - e^{-f})\right) e^{-f(s)}\,, \qquad s \in \mathfrak{S}\,,$$

and putting $f = 0$, we get in particular $1 = -\varphi'(0)$. Hence

$$L_{\xi_s - \delta_s}(f) = -\varphi'\left(E\xi(1 - e^{-f})\right)\,, \qquad s \in \mathfrak{S}\ \text{a.e.}\ E\xi\,,$$

so if $E\xi \in \mathfrak{M}_d$, it follows by (9.1) and Theorem 9.1 that $\xi_s - \delta_s \in M(E\xi, -\varphi')$, $s \in \mathfrak{S}$ a.e. $E\xi$. For general $E\xi$, consider any $\eta \in M(\omega, \varphi)$ with $\omega \in \mathfrak{M}_d$, and conclude as above that $M(\omega, -\varphi')$ exists. Then so does $M(E\xi, -\varphi')$, so the above conclusion remains true by Theorem 3.1.

Conversely, suppose that $\xi_s - \delta_s \overset{d}{=}$ some η, $s \in \mathfrak{S}$ a.e. $E\xi$. To prove that $\xi \in M(E\xi, \varphi)$ for some φ, it suffices by Corollary 9.2 to assume that $E\xi$ is bounded. Letting $n \in N$ be fixed with $P\{\xi\mathfrak{S} = n\} > 0$ and writing τ for a randomly chosen (unit) atom position of ξ, we get by Lemma 11.6 for any $M \in \mathcal{N}$, and for $s \in \mathfrak{S}$ a.e. $E[\xi;\ \xi\mathfrak{S} = n]$

$$P\{\xi - \delta_\tau \in M \mid \xi\mathfrak{S} = n,\ \tau = s\} = P\{\xi_s - \delta_s \in M \mid \xi_s\mathfrak{S} = n\}$$
$$= P\{\xi_s - \delta_s \in M \mid (\xi_s - \delta_s)\,\mathfrak{S} = n - 1\}\,,$$

which shows that $\xi - \delta_\tau$ is conditionally independent of τ, given $\xi\mathfrak{S}$. Letting $\tau_1, \ldots, \tau_n$ be the atom positions of ξ taken in random order, it is then easily seen that τ_i is conditionally independent of $\{\tau_j,\ j \neq i\}$ for all i, and so it follows by induction that $\tau_1, \ldots, \tau_n$ are conditionally independent. Since the τ_j

we obtain

$$Ef(\zeta_s\mathfrak{S}) = \frac{EBR_+ f(BR_+)}{EBR_+}, \quad s \in \mathfrak{S} \text{ a.e. } E\xi, \qquad f \in \mathcal{F}(R_+) . \tag{11.12}$$

Using Lemma 11.8, we may next conclude from (11.11) that the conditional distribution of $(\eta, \zeta) = (\xi_s\{s\},\ \xi_s - \xi_s\{s\}\,\delta_s)$, given that $\xi_s\mathfrak{S} = u$, is a.e. independent of s, and by (11.12) this remains true for the unconditional distribution. Taking expectations in (11.11), it is further seen that (η, ζ) is a.e. symmetrically distributed w.r.t. $E\xi$, and if $\zeta \in S_1(E\xi, \alpha_\zeta, \beta_\zeta)$, we get by (11.8) and (11.11) for any $f \in \mathcal{F}\big(R_+ \times \mathfrak{M}(R_+)\big)$

$$E[f(\eta, B_\zeta) \mid \xi_s\mathfrak{S} = u] = u^{-1}E\left[\textstyle\int f(x, B - x\delta_x)\, B(dx) \mid BR_+ = u\right], \quad \text{a.e.,} \tag{11.13}$$

where $B_\zeta = \alpha_\zeta\delta_0 + \beta_\zeta^1$. Inserting (11.13) into (11.12) yields

$$Ef(\eta, B_\zeta) = \frac{1}{EBR_+}\, E \int f(x, B - x\delta_x)\, B(dx)\,, \quad f \in \mathcal{F}\big(R_+ \times \mathfrak{M}(R_+)\big) . \tag{11.14}$$

Let us now suppose that η and ζ are independent. As before, it suffices to consider the case when $0 < \omega\mathfrak{S} < \infty$, so let us assume that $\xi \in S_1(\omega, \alpha, \beta)$. By (11.14) we obtain for any $t \in R_+$ and $f \in \mathcal{F}\big(\mathfrak{M}(R'_+)\big)$

$$\begin{aligned} Ee^{-\eta t}\, Ef(B_\zeta) = Ee^{-\eta t} f(B_\zeta) &= \frac{1}{EBR_+}\, E \int e^{-xt} f(B - x\delta_x)\, B(dx) \\ &= \frac{1}{EBR_+} \int e^{-xt} Ef(B_x - x\delta_x)\, EB(dx)\,, \end{aligned}$$

so if $0 < Ef(B_\zeta) < \infty$, we get

$$Ee^{-\eta t} = \int e^{-xt}\, \frac{Ef(B_x - x\delta_x)}{Ef(B_\zeta)}\, \frac{EB(dx)}{EBR_+} . \tag{11.15}$$

Since (11.15) holds in particular with $f \equiv 1$, it follows by A 5.1 that

$$Ef(B_x - x\delta_x) = Ef(B_\zeta)\,, \qquad x \in R_+ \text{ a.e. } EB\,, \qquad f \in \mathcal{F}\big(\mathfrak{M}(R'_+)\big) .$$

Turning to an $f \in \mathcal{F}\big(R_+ \times \mathfrak{N}(R'_+)\big)$, we thus obtain in an obvious notation

$$Ef(\alpha_\zeta, \beta_\zeta) = \frac{Ef(\alpha, \beta - \delta_x)\, B(dx)}{EB(dx)} = \frac{Ef(\alpha, \beta - \delta_x)\, \beta(dx)}{E\beta(dx)} = Ef(\alpha_x, \beta_x - \delta_x) \tag{11.16}$$

for any $x > 0$ a.e. $E\beta$, and if $E\alpha > 0$ also

$$Ef(\alpha_\zeta, \beta_\zeta) = \frac{E\alpha f(\alpha, \beta)}{E\alpha} . \tag{11.17}$$

As for ξ above, it suffices in the sequel to consider the restrictions of β to compact subintervals of R'_+, so we may henceforth assume that $E\beta R'_+ < \infty$. Proceeding as in the proof of Theorem 11.5, we may then conclude from (11.16) that, for given $\nu = \beta R'_+$, β is conditionally distributed as a sample process independent of α with intensity $\nu E\beta/E\nu$. From now on we may assume that $E\alpha$ and $E\nu$ are both positive, since otherwise either (i) or (ii) would hold trivially. Combining (11.16) and (11.17), we then obtain for any $t \in R_+$ and $z \in [0, 1]$, and for all $x > 0$ a.e. $E\beta$

$$\frac{E\alpha e^{-\alpha t}z^{\nu}}{E\alpha} = \frac{Ee^{-\alpha t}z^{\nu-1}\beta(dx)}{E\beta(dx)} = \frac{Ee^{-\alpha t}z^{\nu-1}E(\beta(dx) \mid \alpha, \nu)}{E\beta(dx)}$$

$$= \frac{Ee^{-\alpha t}z^{\nu-1}\nu E\beta(dx)/E\nu}{E\beta(dx)} = \frac{Ee^{-\alpha t}\nu z^{\nu-1}}{E\nu},$$

or equivalently

$$Ee^{-\alpha t}\frac{\alpha E(z^{\nu} \mid \alpha)}{E\alpha} = Ee^{-\alpha t}\frac{E(\nu z^{\nu-1} \mid \alpha)}{E\nu}.$$

Considering these expectations as L-transforms in t, it follows by A 5.1 that

$$\frac{\alpha E(z^{\nu} \mid \alpha)}{E\alpha} = \frac{E(\nu z^{\nu-1} \mid \alpha)}{E\nu} \quad \text{a.s.,} \qquad z \in [0, 1], \tag{11.18}$$

and assuming the conditional expectations in (11.18) to be calculated from some family of regular conditional distributions, it is seen that (11.18) may be extended by continuity from any countable dense subset of z-values, and hence that the exceptional null-set in (11.18) may be taken to be independent of z. Writing $\psi_{\alpha}(z) \equiv E(z^{\nu} \mid \alpha)$ and $c = E\nu/E\alpha$, we then get a.s. the differential equations

$$\psi_{\alpha}'(z) = c\alpha\psi_{\alpha}(z), \qquad z \in [0, 1],$$

and since clearly $\psi_{\alpha}(1) \equiv 1$, we obtain a.s. the unique solutions

$$\psi_{\alpha}(z) = e^{-c\alpha(1-z)}, \qquad z \in [0, 1].$$

This means by Table 1 that ν is conditionally a.s. POISSONian with mean $c\alpha$, given α, and hence that β is a COX process directed by $\alpha E\beta/E\alpha$. Proceeding as in the proof of Theorem 9.4, we may thus conclude that $\xi \in S_{\infty}(E\xi, \alpha, \alpha E\beta/E\alpha)$, which shows that (i) is fulfilled.

This proves the necessity of (i) and (ii). The sufficiency of these conditions follows easily by reversing the above arguments, or alternatively, by a direct calculation of L-transforms. □

Notes. Theorem 11.2 was proved by successive generalizations in papers by KERSTAN MATTHES [42], AMBARTZUMJAN [1], MECKE [56], and KUMMER, MATTHES [47], (cf. [44], pp. 116, 206), while Theorems 11.1 and 11.3 are new. The point process case of Theorem 11.4 was proved by MATTHES [55], using entirely different methods, (cf. [44], p. 97). As for Theorem 11.5, special cases were established by SLIVNYAK [74] and PAPANGELOU [66], while the general result was given in [31]. Finally, Theorem 11.7 comes from [37].

Exercises

11.1. As stated above, a random measure ξ with σ-finite intensity is POISSONian iff $\xi_s - \delta_s \overset{d}{=} \xi$, $s \in \mathfrak{S}$ a.e. $E\xi$. (a) Prove this result from Lemma 10.6 and Theorem 11.2. (b) Give a proof in the point process case based on Theorem 11.5. (Hint: Solve the differential equation $\varphi = -\varphi'$, cf. [31].) (c) Show by induction that $\xi_s - \delta_s \overset{d}{=} \xi$ implies $P\{\xi B = n\} \equiv P\{\xi B = 0\}\ (E\xi B)^n/n!$, $n \in Z_+$, and derive the corresponding expression for $P\{\xi B_1 = n_1, \dots, \xi B_k = n_k\}$ when $B_1, \dots, B_k \in \mathcal{B}$ are disjoint. Conclude that ξ must be POISSONian. Cf. [26].)

11.2. Suppose that $\mathfrak{S}$ is countable. Show that there exists some fixed function $f \in \mathcal{F}$, such that a point process ξ on $\mathfrak{S}$ with $E\xi\mathfrak{S} < \infty$ is infinitely divisible iff $P\{\xi = 0\} > 0$ and $P(\xi f)^{-1} | P(\xi_f f)^{-1}$. (Hint: Proceed as in the proof of Theorem 11.1, and refer to Exercise 6.13.)

11.3. Let ξ be a point process on $\mathfrak{S}$ and suppose that $B \in \mathcal{B}$ is such that $P\{\xi \neq 0, \xi B = 0\} = 0$. Show that ξ is then infinitely divisible iff $P\{\xi = 0\} > 0$ and $P\xi^{-1} \mid P\xi_{1_B}^{-1}$. (Hint: For arbitrary $k \in N$ and $B_1, \dots, B_k \in \mathcal{B}$, let $\psi(s, \boldsymbol{t}) = \psi(s, t_1, \dots, t_k)$ be the generating function of $(\xi B, \xi B_1, \dots, \xi B_k)$. Note that $\psi(0, \boldsymbol{t}) \equiv \psi(0, 1)$, and show as in the proof of Theorem 11.3 that $\psi(1, \boldsymbol{t})$ is infinitely divisible.)

11.4. Let ξ be a random measure on $\mathfrak{S}$ with σ-finite intensity. Show that $\xi = \varrho E\xi$ a.s. for some R_+-valued random variable ϱ iff $P\xi_s^{-1}$ is independent of $s \in \mathfrak{S}$ a.e. $E\xi$. (Hint: Let $B_j \in \mathcal{B}$ be arbitrary with $0 < E\xi B_j < \infty$, $j = 1, 2$, and show directly from the definitions that $\xi B_1/E\xi B_1 = \xi B_2/E\xi B_2$ a.s. Then use A 2.2 to show that the P-null set here may be taken to be independent of B_1 and B_2.)

11.5. Let $\xi \in S_\infty(E\xi, \alpha, \gamma)$ where $E\xi \in \mathfrak{M}$ with $E\xi\mathfrak{S} = \infty$, and define $\Gamma = \alpha\delta_0 + \gamma^1$. Further suppose that (η, ζ) is such as in Theorem 11.7, assume that $\zeta \in S_\infty(E\xi, \alpha_\zeta, \gamma_\zeta)$, and put $\Gamma_\zeta = \alpha_\zeta\delta_0 + \gamma_\zeta^1$. Show that

$$Ef(\eta, \Gamma_\zeta) = \frac{1}{E\Gamma R_+} E \int f(x, \Gamma)\, \Gamma(dx), \quad f \in \mathcal{F}(R_+ \times \mathfrak{M}(R_+)) . \tag{11.19}$$

(Cf. [37]. Hint: Prove as in Lemma 10.6 that

$$Ee^{-\xi_s f} = \frac{1}{E\Gamma R_+} Ee^{-\alpha\omega f - \gamma g} \int e^{-xf(s)} \Gamma(dx) , \qquad f \in \mathcal{F} ,$$

where $g(x) \equiv \omega(1 - e^{-xf})$, and show that this implies

$$Ee^{-\eta t - \zeta f} = \frac{1}{E\Gamma R_+} E \int E[e^{-xt - \xi f} | \Gamma]\, \Gamma(dx) , \qquad t \in R_+, f \in \mathcal{F} .$$

Then conclude from Theorem 3.1 that

$$Ef(\eta, \zeta) = \frac{1}{E\Gamma R_+} E \int E[f(x, \xi) \mid \Gamma]\, \Gamma(dx) , \qquad f \in \mathcal{F}(R_+ \times \mathfrak{M}) ,$$

and finally use the fact that Γ and Γ_ζ are obtained by applying the same measurable function to ξ and ζ respectively.)

11.6. Suppose instead that $E\xi\mathfrak{S} < \infty$ in Exercise 11.5. Show that if (η, Γ_ζ), $\Gamma_\zeta = \alpha_\zeta\delta_0 + \gamma_\zeta^1$, is such that $\zeta \in S_\infty(E\xi, \alpha_\zeta, \gamma_\zeta)$ holds conditionally, given (η, Γ_ζ), then $P(\eta, \Gamma_\zeta)^{-1}$ is still unique and (11.19) remains true. (Hint: The uniqueness may be proved as in Corollary 3.2. We may then consider ξ as the restriction to $\mathfrak{S}$ of some symmetrically distributed random measure ξ' with unbounded intensity which is defined on some extended space $\mathfrak{S}'$, and apply (11.19) to ξ'.)

11.7. Prove the last statement of Theorem 11.7 directly from (11.19) in the particular case when $\xi \in$ some $S_\infty(E\xi, \alpha, \gamma)$. (Cf. [37]. Hint: Use A 5.1 and Exercise 11.4.)

11.8. Show that if ξ is such as in (i) of Theorem 11.7, and if $\alpha + \gamma^1 R'_+ = 1$ and $L_\varrho = \varphi$, then $P\eta^{-1} = \alpha\delta_0 + \gamma^1$ while $\zeta \in S_\infty(E\xi, \varrho_\zeta\alpha, \varrho_\zeta\gamma)$ with $L_{\varrho_\zeta} = -\varphi'$. Show also that if ξ is instead such as in (ii) and if $\beta \in M(E\beta, \varphi)$, then $\zeta \in S_1(E\xi, 0, \beta_\zeta)$ with $\beta_\zeta \in M(E\beta, -\varphi')$. (Cf. [37].)

11.9. Show that if ξ, η and ζ are such as in Theorem 11.7, then ζ is independent of η with $\zeta \stackrel{d}{=} \xi$ iff $\xi \in$ some $IA(\alpha\,\omega, \gamma \times \omega)$. (Hint: Apply the results in Exercise 11.8.)

Appendix

Here we collect for easy reference some results outside random measure theory which are needed in the main text. We omit the proofs whenever these are easily available in the literature.

A1. Some basic topology

For any subset B of a topological space, let B°, B^-, B^c and ∂B denote the interior, closure, complement and boundary of B respectively.

A 1.1. *Let B and C be subsets of a topological space $\mathfrak{S}$. Then*

(i) $\partial(B \cup C) \subset \partial B \cup \partial C$,   (ii) $\partial(B \cap C) \subset \partial B \cup \partial C$,   (iii) $\partial(B^c) = \partial B$.

Proof. Since (iii) follows trivially from the fact that $\partial B = B^- \cap (B^c)^-$ while (ii) follows immediately from (i) and (iii), it is enough to prove (i). For this purpose, note that $(B \cup C)^- = B^- \cup C^-$ while $(B \cap C)^- \subset B^- \cap C^-$, and conclude that

$$\partial(B \cup C) = (B \cup C)^- \cap \big((B \cup C)^c\big)^- = (B^- \cup C^-) \cap (B^c \cap C^c)^-$$
$$\subset (B^- \cup C^-) \cap (B^c)^- \cap (C^c)^- = \big(B^- \cap (B^c)^- \cap (C^c)^-\big) \cup \big(C^- \cap (B^c)^- \cap (C^c)^-\big)$$
$$\subset \big(B^- \cap (B^c)^-\big) \cup \big(C^- \cap (C^c)^-\big) = \partial B \cup \partial C. \qquad \square$$

A 1.2. *For elements $x, x_1, x_2, \ldots$ in a topological space, we have $x_n \to x$ iff every sequence $N' \subset N$ contains a subsequence N'' such that $x_n \to x \ (n \in N'')$.* (Cf. [6], p. 16.)

Proof. If $x_n \not\to x$, there must exist some neighbourhood G of x and some subsequence $N' \subset N$ such that $x_n \notin G$, $n \in N'$. But then $x_n \not\to x \ (n \in N'')$ for every further subsequence $N'' \subset N'$. $\qquad \square$

A2. Monotone class theorems

For any class $\mathscr{C}$ of subsets of a fixed set $\mathfrak{S}$, let $\sigma(\mathscr{C})$ denote the σ-algebra *generated* by $\mathscr{C}$, i.e. the smallest σ-algebra in $\mathfrak{S}$ containing $\mathscr{C}$.

A 2.1. *Let $\mathscr{C}$ and $\mathscr{D}$ be classes of subsets of $\mathfrak{S}$ satisfying $\mathscr{C} \subset \mathscr{D}$, and suppose that $\mathscr{C}$ is closed under finite intersections while $\mathscr{D}$ contains $\mathfrak{S}$ and is closed under proper differences and non-decreasing limits. Then $\mathscr{D} \supset \sigma(\mathscr{C})$.* (SIERPIŃSKY 1928, cf. [3].)

For the next result, let $\mathscr{B}$ be a fixed class of subsets of $\mathfrak{S}$, and say that a set $C \subset \mathfrak{S}$ is *bounded* if it may be covered by finitely many sets in $\mathscr{B}$. Given any class $\mathscr{C}$ of bounded sets, let $\hat{\sigma}(\mathscr{C})$ denote the smallest ring which contains $\mathscr{C}$ and is closed under bounded countable unions. (Note that the class of all rings with this property is non-empty since it contains the class of all bounded sets.)

A.2.2. *Let $\mathscr{C}$ and $\mathscr{D}$ be classes of subsets of $\mathfrak{S}$ satisfying $\mathscr{C} \subset \mathscr{D}$, and suppose that $\mathscr{C}$ is a ring of bounded sets while $\mathscr{D}$ is closed under bounded monotone limits. Then $\mathscr{D} \supset \hat{\sigma}(\mathscr{C})$.*

Proof. Proceed as in [20], pp. 27f. (Cf. [59].) $\qquad \square$

we hence obtain

$$\mu\{(s_1, \ldots, s_k): s_1 \leqq x_1, \ldots, s_k \leqq x_k\} = \nu\{(s_1, \ldots, s_k): s_1 \leqq x_1, \ldots, s_k \leqq x_k\}$$

for arbitrary $x_1, \ldots, x_k \in R_+$, which implies $\mu = \nu$. □

For any bounded measures $\mu, \mu_1, \mu_2, \ldots$ on R_+^k, write $\mu_n \xrightarrow{w} \mu$ if $\mu_n f \to \mu f$ for every bounded continuous function $R_+^k \to R$. Note in particular that $\xi_n \xrightarrow{d} \xi$ holds for random vectors $\xi, \xi_1, \xi_2, \ldots$ in R_+^k iff $P\xi_n^{-1} \xrightarrow{w} P\xi^{-1}$.

A 5.2. *Let $k \in N$ and let $\mu_1, \mu_2, \ldots$ be bounded measures on R_+^k. Then $\mu_n \xrightarrow{w}$ some μ iff $\hat{\mu}_n(t) \to$ some $\varphi(t)$, $t \in R_+^k$, where φ is continuous at 0, and in this case $\varphi = \hat{\mu}$. If φ is known to be an L-transform, it is enough to require that $\hat{\mu}_n(t) \to \varphi(t)$ hold for all t in some non-empty open subset of R_+^k.*

Proof. By the very definitions, $\mu_n \xrightarrow{w} \mu$ implies $\hat{\mu}_n(t) \to \hat{\mu}(t)$, $t \in R_+^k$. Conversely, suppose that $\hat{\mu}_n(t) \to \varphi(t)$, $t \in R_+^k$, where φ is continuous at 0. Since $\{\mu_n\}$ is automatically relatively compact when considered as a sequence of measures on the compactified space $R_+^k \cup \{\infty\}$, (cf. A 4.4 or A 7.5), every sequence $N' \subset N$ must contain a subsequence N'' such that $\mu_n \xrightarrow{w}$ some μ on $R_+^k \cup \{\infty\}$ $(n \in N'')$. Now the function $x \to e^{-tx}$ is continuous on $R_+^k \cup \{\infty\}$ whenever $t \in (R_+\backslash\{0\})^k$, so we get $\hat{\mu}_n(t) \to \hat{\mu}(t)$ $(n \in N'')$, $t \in (R_+\backslash\{0\})^k$, and therefore $\hat{\mu}(t) = \varphi(t)$, $t \in (R_+\backslash\{0\})^k$. Moreover, $\hat{\mu}_n(0) = \mu_n R_+^k \to \mu(R_+^k \cup \{\infty\})$ $(n \in N'')$, so as $t \to 0$, $\hat{\mu}(t) = \varphi(t) \to \varphi(0) = \mu(R_+^k \cup \{\infty\})$, which is impossible unless $\mu\{\infty\} = 0$. Thus $\mu_n \xrightarrow{w}$ some μ on R_+^k satisfying $\hat{\mu} = \varphi$ $(n \in N'')$, and since N' was arbitrary, it follows from A 1.2 and A 5.1 that $\mu_n \xrightarrow{w} \mu$ $(n \in N)$. The second statement is proved by combining an argument of this type with the device of analytic continuation. □

We next define the difference operator Δ_h by $\Delta_h f(x) \equiv f(x+h) - f(x)$. Differences of higher order are defined recursively by $\Delta_h^{n+1} = \Delta_h \Delta_h^n$. Given any $a > 0$ and any function $f: [0, a] \to R$, we shall say that f is *absolutely monotone* on $[0, a]$ if $\Delta_h^n f(0) \geqq 0$ for every $n \in N$ and $h \leqq 1/n$. Furthermore, we shall say that a function $f: R_+ \to R$ is *completely monotone* on R_+, if the function $x \to f(a-x)$ is absolutely monotone on $[0, a]$ for every $a > 0$, i.e. if

$$(-1)^n \Delta_h^n f(x) \geqq 0\,, \qquad x, h > 0\,, \qquad n \in N\,.$$

A 5.3 (BERNSTEIN's theorem, cf. [15], pp. 223, 439).

A function f: $[0, 1] \to R$ with $f(1) = 1$ is the probability generating function of some Z_+-valued random variable iff f is absolutely monotone and continuous on $[0, 1]$.

A function f: $R_+ \to R$ with $f(0) = 1$ is the L-transform of some R_+-valued random variable iff f is completely monotone and continuous on R_+.

A6. Locally compact spaces

Throughout A 6 and A 7, $\mathfrak{S}$ is assumed to be a *locally compact second countable Hausdorff space*. This means that every point in $\mathfrak{S}$ has a compact neighbourhood, that $\mathfrak{S}$ has a countable base and that distinct points may be separated by disjoint neighbourhoods. In this case, $\mathfrak{S}$ is known to be *Polish*, i.e. there exists some separable and complete metrization ϱ of $\mathfrak{S}$, (cf. [73]).

A set $B \subset \mathfrak{S}$ is said to be (topologically) *bounded* or *relatively compact*, if its closure B^- is compact. (Note that a sphere w.r.t. ϱ above need not be bounded in this sense.) Write $\mathcal{F}_c$ for the class of continuous functions $\mathfrak{S} \to R_+$ with compact support.

A 6.1. *Given any compact (or open) set $B \subset \mathfrak{S}$, there exist some compact (or open bounded, respectively) sets $B_n \subset \mathfrak{S}$ and some $f_n \in \mathcal{F}_c$, $n \in N$, such that*

$$1_B \leqq f_n \leqq 1_{B_n} \downarrow 1_B \qquad (\text{or } 1_B \geqq f_n \geqq 1_{B_n} \uparrow 1_B) .$$

Proof. Consider a countable base consisting of bounded sets, and let G_n be the union of the first n basic sets. Then $G_n \uparrow \mathfrak{S}$, and since each G_n is open, any fixed bounded set B is covered by some G_n. If B is compact, then $\varrho(B, G_n^c)$ must be positive for this G_n, and so $B^\varepsilon = \{s \in \mathfrak{S}: \varrho(s, B) \leq \varepsilon\}$ is bounded for sufficiently small $\varepsilon > 0$. We may thus take $B_n = B^{1/n}$ for large $n \in N$ and define

$$f_n(s) = 1 - n\{\varrho(s, B) \wedge n^{-1}\} , \qquad s \in \mathfrak{S}, \qquad n \in N .$$

The proof for open bounded B is similar. If B is open but unbounded, we may apply the same argument to the sequence $B \cap G_n$, $n \in N$. □

A7. Measures on locally compact spaces

For $\mathfrak{S}$ as in A 6, let $\mathcal{S}$ denote the Borel algebra in $\mathfrak{S}$ und let $\mathcal{B}$ be the ring of all bounded $\mathcal{S}$-sets. Further denote by $\mathfrak{M}$ the class of all Radon measures on $(\mathfrak{S}, \mathcal{B})$, i.e. of all measures μ such that $\mu B < \infty$ for all $B \in \mathcal{B}$. Let $\mathfrak{N}$ be the subspace of all Z_+-valued measures in $\mathfrak{M}$. Write $f\mu$ for the measure $(f\mu)B = \int_B f(s)\, \mu(ds)$, $B \in \mathcal{S}$, and put $B\mu = 1_B\mu$, $\mu f = (f\mu)\mathfrak{S}$. Further define, for fixed $\mu \in \mathfrak{M}$, $\mathcal{B}_\mu = \{B \in \mathcal{B}: \mu\partial B = 0\}$.

The class of all finite intersections of $\mathfrak{M}$-sets (or $\mathfrak{N}$-sets) of the form $\{\mu: s < \mu f < t\}$ with arbitrary $f \in \mathcal{F}_c$ and $s, t \in R$ may serve as a base for a topology on $\mathfrak{M}$ (or $\mathfrak{N}$), to be called the *vague* one. Note that μ_n tends to μ in this topology (or *vaguely*; written $\mu_n \xrightarrow{v} \mu$), iff $\mu_n f \to \mu f$ for every $f \in \mathcal{F}_c$. When considering the subspace of all bounded measures in $\mathfrak{M}$ (or $\mathfrak{N}$), we may replace $\mathcal{F}_c$ in the above definition by the class of all bounded continuous functions $\mathfrak{S} \to R_+$, thus obtaining the *weak* topology with the property that μ_n tends *weakly* to μ ($\mu_n \xrightarrow{w} \mu$) iff $\mu_n f \to \mu f$ for all bounded continuous f. Note in particular that, for random elements $\xi, \xi_1, \xi_2, \ldots$ in $(\mathfrak{S}, \mathcal{S})$, $\xi_n \xrightarrow{d} \xi$ iff $P\xi_n^{-1} \xrightarrow{w} P\xi^{-1}$. We shall combine this fact with the following lemma to derive properties of the vague topology from results in A 4.

A 7.1. *Suppose that $1 \geqq f_k \geqq 1_{G_k} \uparrow 1$ for some open sets $G_k \in \mathcal{B}$ and some $f_k \in \mathcal{F}_c$, $k \in N$,* (cf. A 6.1). *Then $\mu_n \xrightarrow{v}$ some μ iff $f_k\mu_n \xrightarrow{w}$ some μ_k' for each $k \in N$, and in that case $\mu_k' = f_k\mu$, $k \in N$.*

Proof. Suppose that $f_k\mu_n \xrightarrow{w} \mu_k'$, $k \in N$. For fixed $k \in N$, we have $G_k\mu_k' = G_k\mu_{k+1}' = \cdots = \mu_k''$, say. In fact, $\mu_k' f = \mu_{k+1}' f = \cdots$ for all $f \in \mathcal{F}_c$ with support in G_k, so using A 6.1 and monotone convergence, it follows that $\mu_k' B = \mu_{k+1}' B = \cdots$ for all open subsets of G_k. By A 2.1, the last fact extends to arbitrary $B \in B \cap G_k$.

It is now easy to verify that the relations $\mu B = \max_k \mu_k'' B$, $B \in \mathscr{B}$, define a measure $\mu \in \mathfrak{M}$ satisfying $\mu_n \xrightarrow{v} \mu$. The assertion in the converse direction is obvious. □

A 7.2. *Let* $\mu, \mu_1, \mu_2, \ldots \in \mathfrak{M}$. *Then the following statements are equivalent.* (Cf. [44], p. 222.)

(i) $\mu_n \xrightarrow{v} \mu$,
(ii) $\mu_n B \to \mu B$ *for all* $B \in \mathscr{B}_\mu$,
(iii) $\limsup_{n\to\infty} \mu_n F \leqq \mu F$ *and* $\liminf_{n\to\infty} \mu_n G \geqq \mu G$ *for all closed* $F \in \mathscr{B}$ *and open* $G \in \mathscr{B}$.

Proof. In view of A 4.1 and A 7.1, (i) implies (ii) and (iii), and conversely, (ii) follows trivially from (iii). It remains to prove that (ii) implies (i), so let us assume that (ii) is true. For any fixed $f \in \mathscr{F}_c$, choose a set $C \in \mathscr{B}_\mu$ containing the support F of f. (For the existence, note that $F_\varepsilon = \{s\colon \varrho(s, F) \leqq \varepsilon\}$ is bounded for small $\varepsilon > 0$, say for $\varepsilon \in (0, c)$, (cf. A 6.1), and that $\partial F_\varepsilon \subset \{s\colon \varrho(s, F) = \varepsilon\}$. Thus $\mu \partial F_\varepsilon$ can be positive for at most countably many $\varepsilon \in (0, c)$.) By A 1.1, $B \cap C \in \mathscr{B}_\mu$ whenever $B \in \mathscr{B}_\mu$, so for such a B, (ii) yields $(C\mu_n)B = \mu_n(B \cap C) \to \mu(B \cap C) = (C\mu)B$, and it follows easily from A 4.1 that $C\mu_n \xrightarrow{w} C\mu$. In particular, $\mu_n f = (C\mu_n)f \to (C_\mu)f = \mu f$, which proves (i). □

A.7.3. *Let* $\mu_0, \mu_1, \ldots \in \mathfrak{M}$ *with* $\mu_n \xrightarrow{v} \mu_0$, *and let* $B_0, B_1, \ldots \in \mathscr{S}$ *be such that* $\mu_n B_n^c = 0$, $n \in Z_+$. *Further suppose that* $f_0, f_1, \ldots$ *are uniformly bounded measurable functions* $\mathfrak{S} \to R_+$ *with uniformly bounded supports, such that* $f_n(s_n) \to f_0(s_0)$ *whenever* $s_n \in B_n$, $n \in Z_+$, *with* $s_n \to s_0$. *Then* $\mu_n f_n \to \mu_0 f_0$. *In particular,* $\mu_n f \to \mu_0 f$ *for every bounded measurable function* $f\colon \mathfrak{S} \to R_+$ *with bounded support satisfying* $\mu_0 D_f = 0$.

Proof. Since the f_n have uniformly bounded supports, we may assume by A 7.1 that $\mu_n \xrightarrow{w} \mu_0$, and excluding the trivial case $\mu_0 = 0$, we may further assume the μ_n to be normalized, i.e. to be probability measures on $\mathfrak{S}$. But then A 4.2 yields $\mu_n f_n^{-1} \xrightarrow{w} \mu_0 f_0^{-1}$, and the f_n being uniformly bounded, it follows by A 4.3 that $\mu_n f_n \to \mu_0 f_0$. □

A 7.4. $\mathfrak{N}$ *is a vaguely closed subset of* $\mathfrak{M}$. (Cf. [27], p. 200, and [44], p. 240.)

Proof. Let $\mu_1, \mu_2, \ldots \in \mathfrak{N}$ and suppose that $\mu_n \xrightarrow{v}$ some $\mu \in \mathfrak{M}$. Then A 7.2 yields $\mu B \in Z_+$ for every $B \in \mathscr{B}_\mu$. Since $\mathscr{B}_\mu$ is a ring (cf. A 1.1) satisfying $\hat{\sigma}(\mathscr{B}_\mu) = \mathscr{B}$, (cf. the proof of A 7.2), we may apply A 2.2 to conclude that $\mu B \in Z_+$ for all $B \in \mathscr{B}$. □

A 7.5 (cf. [3]). *A subset* M *of* $\mathfrak{M}$ *or* $\mathfrak{N}$ *is relatively compact in the vague topology iff*

$$\sup_{\mu \in M} \mu B < \infty\,, \qquad B \in \mathscr{B}\,,$$

and in the weak topology iff

$$\sup_{\mu \in M} \mu \mathfrak{S} < \infty \quad \text{and} \quad \inf_{B \in \mathscr{B}} \sup_{\mu \in M} \mu B^c = 0\,.$$

Proof. By A 7.4, it suffices to consider $\mathfrak{M}$. In this case, the assertion for the weak topology follows directly from A 4.4, and then we obtain the one for the vague topology by application of A 7.1. □

A 7.6. *For bounded* $\mu, \mu_1, \mu_2, \ldots \in \mathfrak{M}$, *the following statements are equivalent.*

(i) $\mu_n \xrightarrow{w} \mu$,

(ii) $\mu_n \xrightarrow{v} \mu$ *and* $\mu_n \mathfrak{S} \to \mu \mathfrak{S}$,

(iii) $\mu_n \xrightarrow{v} \mu$ *and* $\inf_{B \in \mathcal{B}} \limsup_{n \to \infty} \mu_n B^c = 0$.

Proof. The implication (i) $\Rightarrow$ (ii) is obvious. We next assume that (ii) holds and let $\varepsilon > 0$ be arbitrary. Choose an open set $G \in \mathcal{B}$ with $\mu G^c < \varepsilon/2$ (cf. A 6.1), and let $f \in \mathcal{F}_c$ be such that $f \leqq 1_G$ while $\mu(1_G - f) < \varepsilon/2$ (cf. A 6.3). Then

$$\limsup_{n \to \infty} \mu_n G^c \leqq \limsup_{n \to \infty} (\mu_n \mathfrak{S} - \mu_n f) = \mu \mathfrak{S} - \mu f = \mu G^c + \mu(1_G - f) < \varepsilon ,$$

proving (iii). Finally, (iii) implies by A 7.5 that $\{\mu_n\}$ is weakly relatively compact, and so any sequence $N' \subset N$ contains a subsequence N'' such that $\mu_n \xrightarrow{w}$ some μ' $(n \in N'')$. But then $\mu_n \xrightarrow{v} \mu'$ $(n \in N'')$, so we get $\mu = \mu'$, and (i) follows by A 1.2. □

A 7.7. $\mathfrak{M}$ *and* $\mathfrak{N}$ *are Polish in the vague topologies.* (Cf. [7], p. 61.)

Proof. By A 7.4, it is enough to consider $\mathfrak{M}$. Let $\mathcal{C}$ be a countable base in $\mathfrak{S}$ and assume without loss that $\mathcal{C}$ is closed under finite unions and that $\mathcal{C} \subset \mathcal{B}$. For every $C \in \mathcal{C}$, there exists by A 6.1 a sequence $f_{C1}, f_{C2}, \ldots \in \mathcal{F}_c$ with $f_{Cn} \uparrow C$. Let $f_1, f_2, \ldots$ be an enumeration of all the f_{Cn}, $C \in \mathcal{C}$, $n \in N$. Then every $\mu \in \mathfrak{M}$ is uniquely determined by $\{\mu f_k, k \in N\}$. In fact, the latter sequence determines μC, $C \in \mathcal{C}$, by monotone convergence, and so the uniqueness of μ follows by applying A 2.1 to $B \cap C$ for arbitrary $C \in \mathcal{C}$.

Next observe that $\mu_n \xrightarrow{v}$ some μ iff $\mu_k f_k \to$ some $c_k \in R_+$ for all $k \in N$, and that in this case $c_k \equiv \mu f_k$. In fact, suppose that $\mu_k f_k \to c_k$, $k \in N$. Then $\{\mu_k\}$ is relatively compact by A 7.5, so any sequence $N' \subset N$ must contain a subsequence N'' such that $\mu_n \xrightarrow{v}$ some μ $(n \in N'')$. But then $c_k \equiv \mu f_k$, and it follows as above that μ is unique. Thus $\mu_n \xrightarrow{v} \mu$ by A 1.2.

According to this criterion for vague convergence, the vague topology in $\mathfrak{M}$ is induced by the metric

$$\varrho_{\mathfrak{M}}(\mu, \mu') = \sum_{k=1}^{\infty} 2^{-k} \left[1 - \exp\left(|\mu f_k - \mu' f_k|\right)\right] , \qquad \mu, \mu' \in \mathfrak{M} ,$$

and it is easy to verify that this metrization is separable and complete. □

We finally point out that the subspace of bounded measures in $\mathfrak{M}$ or $\mathfrak{N}$ is even Polish in the weak topology, (cf. Prohorov [68], p. 167). However, this fact is not needed here.

Exercises

1. Show that addition and multiplication by scalars are vaguely continuous operations in $\mathfrak{M}$, while convolution is a weakly continuous operation in the subspace of bounded measures. (Hint: Cf. [6], p. 21, for the last assertion.)

2. Let $\mu, \mu_1, \mu_2, \ldots \in \mathfrak{M}$ and let $\mathcal{J} \subset \mathcal{B}_\mu$ be a DC-semiring, (cf. Section 1). Prove that $\mu_n \xrightarrow{v} \mu$ iff $\mu_n I \to \mu I$, $I \in \mathcal{J}$. (Hint: Use A 1.2, A 2.1, A 7.2, A 7.5 and Lemma 1.1.)

3. Show that the sets $\{\mu: \mu F < t\}$ and $\{\mu: \mu G > t\}$ are vaguely open in both $\mathfrak{M}$ and $\mathfrak{N}$ for all closed $F \in \mathcal{B}$ and open $G \in \mathcal{B}$, and for $t \in R$. (Cf. [31]. Hint: Use condition (iii) in A 7.2.)

4. Let $b > 0$, $b_1, b_2, \ldots \in R_+$ and $t, t_1, t_2, \ldots \in \mathfrak{S}$. Show that $b_n \delta_{t_n} \xrightarrow{v} b\delta_t$ iff $b_n \to b$ and $t_n \to t$. (Hint: Use the results of Exercise 3.)

5. Let $\mu, \mu_1, \mu_2, \ldots \in \mathfrak{M}$ with $\mu_n \xrightarrow{v} \mu$, and let $B \in \mathcal{B}_\mu$. Show that $B\mu_n \xrightarrow{w} B\mu$.

6. Let $\{\mu_n\}$ and $\{f_n\}$ be such as in A 7.3, except that we neither assume the functions f_n themselves nor their supports to be bounded. Suppose instead that

$$\inf_{B \in \mathcal{B}} \limsup_{n \to \infty} (f_n \mu_n) B^c = 0 \quad \text{and} \quad \lim_{r \to \infty} \limsup_{n \to \infty} \int_{\{f_n(s) > r\}} f_n(s)\, \mu_n(ds) = 0 .$$

Prove that the conclusion of A 7.3 remains true. (Hint: Approximate the μ_n as in A 7.1 by measures with uniformly bounded supports and the f_n by uniformly bounded functions. Then apply A 7.3. The finiteness of $\mu_0 f_0$ may be proved in the same way as FATOU's lemma.)

7. Use A 7.1 and the fact quoted at the end of A7 to give a new proof of A 7.7.

8. Prove A 7.6 directly, without referring to A 7.5.

Exercises added in proof

3.14. Let p $\in (0,1]$, and let ξ be a p-thinning of η. Show that ξ and η are simultaneously Coxian. (Hint: Use Corollary 3.2.)

3.15. Let ξ and η be simple point processes (or diffuse random measures) on $\mathfrak{S}$, let $\mathcal{U} \subset \mathcal{B}$ be a DC-ring and let $t \in (0, \infty]$ (or $t \in (0, \infty)$ respectively) be fixed. Show that ξ and η are independent iff

$$Ee^{-t(\xi U + \eta V)} = Ee^{-t\xi U}\, Ee^{-t\eta V}, \quad U, V \in \mathcal{U} .$$

(Hint: Proceed as in the proofs of Theorems 3.3 and 3.4.)

7.15. Let ξ be a simple point process on $\mathfrak{S}$ without any fixed atoms, let $\mathcal{U} \subset \mathcal{B}$ be a DC-ring and let $t \in (0, \infty]$ be fixed. Show that ξ is POISSONian iff

$$Ee^{-t\xi(U \cup V)} = Ee^{-t\xi U}\, Ee^{-t\xi V}, \quad U, V \in \mathcal{U} \text{ with } U \cap V = \emptyset.$$

(Hint: Show that a measure $\lambda \in \mathfrak{M}_d$ may be defined by $\lambda U = -\log Ee^{-t\xi U}$, $U \in \mathcal{U}$, and apply Theorem 3.3 or 3.4. Alternatively, the proof may be based on Exercise 3.15 and Corollary 7.4.) State and prove the corresponding result for diffuse random measures.

7.16. Let ξ be a diffuse random measure on $\mathfrak{S}$ and let $\mathcal{J} \subset \mathcal{B}$ be a DC-semiring. Show that ξ is a. s. non-random iff ξI and ξJ are uncorrelated for disjoint $I, J \in \mathcal{J}$. (Hint: Calculate $E(\xi I)^2$ for arbitrary $I \in \mathcal{J}$ by considering a null-array of partitions of I. Cf. DAVIDSON [10] and KRICKEBERG [46a].) Show also that the corresponding result for simple point processes is false, i. e. that there exist non-POISSONian simple point processes without any fixed atoms and with uncorrelated increments. (Hint: Consider a non-POISSONian mixed sample process with mixing variable ν satisfying $E\nu^2 = (E\nu)^2 + E\nu$. Independent processes of this kind may be combined to yield examples with infinitely many atoms.)

7.17. Let $\{\xi_{nj}\}$ be a null-array of random measures on $\mathfrak{S}$, let $\lambda \in \mathfrak{M}$ be non-random and let $\mathcal{J} \subset \mathcal{B}_\lambda$ be a DC-semiring. Show that $\sum_j \xi_{nj} \xrightarrow{d} \lambda$ iff

$$\sum_j E[\xi_{nj} I : \xi_{nj} I < 1] \to \lambda I \quad \text{and} \quad \sum_j P\{\xi_{nj} I > \varepsilon\} \to 0, \quad \varepsilon > 0 ,$$

for all I $\in \mathcal{J}$. (Note the similarity with Corollary 7.5. Hint: Extend the above relations to the ring generated by $\mathcal{J}$ and apply Theorem 7.2.)

8.7. Show that Theorem 8.4 remains true with the p_n interpreted as functions $\mathfrak{S} \to (0,1]$, provided the convergence $p_n \to 0$ is taken to be uniform on bounded sets.

8.8. For $n, j \in N$, let ξ_{nj} be a p_{nj}-thinning of η_{nj}, where the η_{nj} are independent point processes on $\mathfrak{S}$ while $p_{nj} \in \mathcal{F}$ with $\sup_j \sup_{s \in B} p_{nj}(s) \to 0$, $B \in \mathcal{B}$. Show that $\sum_j \xi_{nj} \xrightarrow{d}$ some ξ iff $\sum_i p_{nj}\eta_{nj} \xrightarrow{d}$ some η, and that ξ is then a Cox process directed by η. Assuming $\{p_{nj}\eta_{nj}\}$ to be a null-array, when does $\sum_j \xi_{nj}$ converge to a Poisson process with intensity $\lambda \in \mathfrak{M}$? (Hint: Make estimates in the L-transforms based on the fact that $\log(1 + x) \sim x$ as $x \to 0$. For the Poissonian case, refer to Exercise 7.17 above. Cf. Fichtner [15a] for a first step in the Poissonian case.)

References

[1] Ambartzumjan, R. V.: On an equation for stationary point processes. (In Russian.) *Dokl. Akad. Nauk Armjanskoj SSR* **42** (1966), 141–147.

[2] —: Palm distributions and superpositions of independent point processes in R^n, In *Stochastic Point Processes: Statistical Analysis, Theory, and Applications*, pp. 626–645. Wiley-Interscience, New York 1972.

[3] Bauer, H.: *Probability Theory and Elements of Measure Theory*. Holt, Rinehart and Winston, New York 1972.

[4] Belyaev, Yu. K.: Limit theorems for dissipative flows. *Theor. Probab. Appl.* 8 (1963), 165–173.

[5] —: Elements of the general theory of random streams. (In Russian.) Appendix 2 to the Russian ed. of Cramér, H. and Leadbetter, M. R., *Stationary and Related Stochastic Processes*. MIR, Moskow 1969.

[6] Billingsley, P.: *Convergence of Probability Measures*. Wiley, New York 1968.

[7] Bourbaki, N.: *Intégration*, 1re éd., Chaps. 1–4. Hermann, Paris 1952.

[8] Bühlmann, H.: Austauschbare stochastische Variabeln und ihre Grenzwertsätze. *Univ. California Publ. Statist.* **3** (1960), 1–36.

[9] Daley, D. J.: Various concepts of orderliness for point processes. In *Stochastic Geometry*, pp. 148–161. Wiley, New York 1974.

[10] Davidson, R.: Stochastic processes of flats and exchangeability. In *Stochastic Geometry*, pp. 13–45, Wiley, New York 1974.

[11] Debes, H., Kerstan, J., Liemant, *A.* and Matthes, K.: Verallgemeinerungen eines Satzes von Dobrushin I. *Math. Nachr.* 47 (1970), 183–244.

[12] Dobrushin, R. L.: On Poissson laws for distributions of particles in space. (In Russian.) *Ukrain. Mat. Z.* 8 (1956), 127–134.

[13] Doeblin, W.: Sur les sommes d'un grand nombres de variables aléatoires indépendantes. *Bull. Soc. Math. France* **53** (1939), 23–32, 35–64.

[14] Doob, J. L.: *Stochastic Processes*. Wiley, New York 1953.

[15] Feller, W.: *An Introduction to Probability Theory and its Applications* II, 2nd ed. Wiley, New York 1971.

[15a] Fichtner, K. H.: Schwache Konvergenz von unabhängigen Überlagerungen verdünnter zufälliger Punktfolgen. *Math. Nachr.* **66** (1975), 333–341.

[16] Gihman, I. I. and Skorohod, A. V.: *Introduction to the Theory of Random Processes*. Saunders, Philadelphia 1969.

[17] Goldman, J. R.: Stochastic point processes: limit theorems. *Ann. Math. Statist.* **38** (1967), 771–779.

[18] Grandell, J.: A note on characterization and convergence of non-atomic random measures. In *Abstract of Communications T. 1.*, pp. 175–176. International Conference on Probability Theory and Mathematical Statistics at Vilnius 1973.

[19] Grigelionis, B.: On the convergence of sums of random step processes to a Poisson process. *Teor. Probab. Appl.* 8 (1963), 172–182.

[20] HALMOS, P. R.: *Measure Theory*. Van Nostrand, New York 1950.

[21] HARRIS, T. E.: *The Theory of Branching Processes*. Springer, Berlin 1963.

[22] —: Counting measures, monotone random set functions. *Z. Wahrscheinlichkeitstheorie verw. Gebiete* **10** (1968), 102—119.

[23] —: Random measures and motions of point processes. *Z. Wahrscheinlichkeitstheorie verw. Gebiete* **18** (1971), 85—115.

[24] HINČIN, A.: *Mathematical Methods in the Theory of Queueing*. Griffin, London 1960. (The Russian original appeared in 1955.)

[25] JAGERS, P.: On the weak convergence of superpositions of point processes. *Z. Wahrscheinlichkeitstheorie verw. Gebiete* **22** (1972), 1—7.

[26] —: On Palm probabilities. *Z. Wahrscheinlichkeitstheorie verw. Gebiete* **26** (1973), 17—32.

[27] —: Aspects of random measures and point processes. In *Advances in Probability and Related Topics* **3**, pp. 179—239. Marcel Decker, New York 1974.

[28] JIŘINA, M.: Branching processes with measure-valued states. In *Transactions of the third Prague Conference on Information Theory, Statistical Decision Functions, Random processes*, pp. 333—357. Prague 1964.

[29] —: Asymptotic behaviour of measure-valued branching processes. *Rozpravy Československé Akad. Věd, Řada Mat. a Přírod. Věd* **76**, No. 3, Praha 1966.

[30] KALLENBERG, O.: Characterization and convergence of random measures and point processes. Thesis. Dept. of Mathematics, Chalmers Inst. of Technology, Göteborg 1972.

[31] —: Characterization and convergence of random measures and point processes. *Z. Wahrscheinlichkeitstheorie verw. Gebiete* **27** (1973), 9—21.

[32] —: A canonical representation of symmetrically distributed random measures. In *Mathematics and Statistics, Essays in Honour of Harald Bergström*, pp. 41—48. Teknologtryck, Göteborg 1973.

[33] —: Canonical representations and convergence criteria for processes with interchangeable increments. *Z. Wahrscheinlichkeitstheorie verw. Gebiete* **27** (1973), 23—36.

[34] —: Conditions for continuity of random processes without discontinuities of the second kind. *Ann. Probab.* **1** (1973), 519—526.

[35] —: Characterization of continuous random processes and signed measures. *Studia Sci. Math. Hungar.* **8** (1973), 473—477.

[36] —: Extremality of Poisson and sample processes. *J. Stoch. Proc. Appl.* **2** (1974), 73—83.

[37] —: On symmetrically distributed random measures. *Trans. Amer. Math. Soc.* **202** (1975), 105—121.

[38] —: Limits of compound and thinned point processes. *J. Appl. Probab.* **12** (1975), 269—278.

[39] —: Infinitely divisible processes with interchangeable increments and random measures under convolution. *Z. Wahrscheinlichkeitstheorie verw. Gebiete* **32** (1975), 309—321.

[40] KENDALL, D. G.: Foundations of a theory of random sets. In *Stochastic Geometry*, pp. 322—376. Wiley, New York 1974.

[41] KERSTAN, J. and MATTHES, K.: Stationäre zufällige Punktfolgen II. *Jahresbericht der DMV* **66** (1964), 106—118.

[42] —: Verallgemeinerung eines Satzes von Sliwnjak. *Rev. Roumaine Math. Pures Appl.* **9** (1964), 811—829.

[43] —: A generalization of the Palm-Hinčin theorem. (In Russian.) *Ukrain. Mat. Z.* **17** (1965), 29—36.

[44] KERSTAN, J., MATTHES, K. and MECKE, J.: *Unbegrenzt teilbare Punktprozesse*. Akademie-Verlag, Berlin 1974.

[45] KINGMAN, J. F. C.: Completely random measures. *Pacific J. Math.* **21** (1967), 59—78.

[46] KRICKEBERG, K.: The Cox process. *Inst. Nat. Alta Mat., Symp. Math.* **9** (1972), 151—167.

[46a] —: Moments of point processes. In *Stochastic Geometry*, pp. 89—113. Wiley, New York 1974.

[47] Kummer, G. and Matthes, K.: Verallgemeinerung eines Satzes von Sliwnjak II—III. *Rev. Roumaine Math. Pures Appl.* **15** (1970), 845—870, 1631—1642.

[48] Kurtz, T. G.: Point processes and completely monotone set functions. *Z. Wahrscheinlichkeitstheorie verw. Gebiete* **31** (1974), 57—67.

[49] Leadbetter, M. R.: On basic results of point process theory. In *Proc. 6th Berkeley Symp. Math. Statist. Probab.* **3**, pp. 449—462. Univ. of California Press, Berkeley 1972.

[50] Lee, P. M.: Infinitely divisible stochastic processes. *Z. Wahrscheinlichkeitstheorie verw. Gebiete* **7** (1967), 147—160.

[51] —: Some examples of infinitely divisible point processes. *Studia Sci. Math. Hungar.* **3** (1968), 219—224.

[52] Loève, M.: *Probability Theory*, 3rd. ed. Van Nostrand, Princeton 1963.

[53] Lukacs, E.: *Characteristic Functions*, 2nd ed. Griffin, London 1970.

[54] Matthes, K.: Unbeschränkt teilbare Verteilungsgesetze stationärer zufälliger Punkt folgen. *Wiss. Z. Hochsch. Elektro. Ilmenau* **9** (1963), 235—238.

[55] —: Eine Charakterisierung der kontinuierlichen unbegrenzt teilbaren Verteilungsgesetze zufälliger Punktfolgen. *Revue Roumaine Math. Pures Appl.* **14** (1969), 1121—1127.

[56] Mecke, J.: Stationäre zufällige Maße auf lokalkompakten Abelschen Gruppen. *Z. Wahrscheinlichkeitstheorie verw. Gebiete* **9** (1967), 36—58.

[57] —: Eine charakteristische Eigenschaft der doppelt stochastischen Poissonschen Prozesse. *Z. Wahrscheinlichkeitstheorie verw. Gebiete* **11** (1968), 74—81.

[58] —: Zufällige Maße auf lokalkompakten Hausdorffschen Räumen. *Beiträge zur Analysis* **3** (1972), 7—30.

[59] Mönch, G.: Verallgemeinerung eines Satzes von A. Rényi. *Studia Sci. Math. Hungar.* **6** (1971), 81—90.

[60] Moran, P. A. P.: A non-Markovian quasi-Poisson process. *Studia Sci. Math. Hungar.* **2** (1967), 425—429.

[61] Moyal, J. E.: The general theory of stochastic population processes. *Acta Math.* **108** (1962), 1—31.

[62] Nawrotzki, K.: Ein Grenzwertsatz für homogene zufällige Punktfolgen (Verallgemeinerung eines Satzes von A. Rényi). *Math. Nachr.* **24** (1962), 201—217.

[63] —: Mischungseigenschaften stationärer unbegrenzt teilbarer Maße. *Math. Nachr.* **38** (1968), 97—114.

[64] Ososkov, G. A.: A limit theorem for flows of homogeneous events. *Theor. Probab. Appl.* **1** (1956), 248—255.

[65] Palm, C.: Intensitätsschwankungen in Fernsprechverkehr. *Ericsson Technics* **44** (1943), 1—189.

[66] Papangelou, F.: On the Palm probabilities of processes of points and processes of lines. In *Stochastic Geometry*, pp. 114—147. Wiley, New York 1974.

[67] Prekopa, A.: On secondary processes generated by a random point distribution of Poisson type. *Ann. Univ. Sci. Budapest. Sect. Math.* **1** (1958), 153—170.

[68] Prohorov, Yu. V.: Convergence of random processes and limit theorems in probability theory. *Theor. Probab. Appl.* **1** (1956), 157—214.

[69] —: Random measures on a compactum. *Soviet Math. Dokl.* **2** (1961), 539—541.

[70] Rényi, A.: A characterization of Poisson processes. (In Hungarian with summaries in Russian and English.) *Magyar Tud. Akad. Mat. Kutato Int. Közl.* **1** (1956), 519—527.

[71] —: Remarks on the Poisson process. *Studia Sci. Math. Hungar.* **2** (1967), 119—123.

[72] Ryll-Nardzewski, C.: Remarks on processes of calls. In *Proc. 4th Berkeley Symp. Math. Statist. Probab.* **2**, pp. 455—465. Univ. of California Press, Berkeley 1961.

[73] Simmons, G. F.: *Introduction to Topology and Modern Analysis.* McGraw-Hill, New York 1963.

[74] SLIVNYAK, I. M.: Some properties of stationary flows of homogeneous random events. *Theor. Probab. Appl.* **7** (1962), 336—341, **9** (1964), 168.
[75] SZÁSZ, D. O. H.: Once more on the Poisson process. *Studia Sci. Math. Hungar.* **5** (1970), 441—444.
[76] WALDENFELS, W. v.: Characteristische Funktionale zufälliger Maße. *Z. Wahrscheinlichkeitstheorie verw. Gebiete* **10** (1968), 279—283.
[77] WESTCOTT, M.: Some remarks on a property of the Poisson process. *Sankhyā, Ser. A*, **35** (1973), 29—34.

INDICES

Authors

Notation

Terminology

Some scattered topics